Military Modeling

Wayne P. Hughes, Jr., Editor

John A. Battilega
Thomas A. Brown
Marion R. Bryson
Stephan M. Drezner
John A. Englund
John Friel
Judith K. Grange
Robert A. Hallex
Richard J. Hillestad
Stephen W. Leibholz
Alfred Lieberman
James J. Martin
Wilbur B. Payne
Bernard D. Rostker
Clayton J. Thomas

ISBN 0-930473-00-0. Library of Congress catalog card number 84-61171. Printed in the United States of America.

The Military Operations Research Society

This monograph is the fourth book publication of the Military Operations Research Society, the professional society of the community of practitioners of operations research in the context of national security.

The purpose of the Military Operations Research Society is to enhance the quality and effectiveness of military operations research. To accomplish this purpose, the Society provides media for professional exchange and peer criticism among students, theoreticians, practitioners, and users of military operations research. These media consist primarily of the annual MORS symposia, their published proceedings, and special-purpose monographs. The forum provided by these media is directed to display the state of the art, to encourage consistent professional quality, to stimulate communication and interaction between practitioners and users, and to foster the interest and development of students of operations research. In performing its function, the Military Operations Research Society does not make or advocate official policy nor does it attempt to influence the formulation of policy.

The Military Operations Research Society is operated by a Board of Directors consisting of 28 members, each of whom is elected by vote of the Board to serve a term of four years. Nominees for this election are individuals who have attained recognition and prominence in the field of military operations research and who have demonstrated an active interest in its programs and activities. Since a major portion of the Society's affairs is connected with classified services to military sponsors, the Society does not have a general membership in the sense that other professional societies have them. In place of this, the Society maintains a general distribution list of its clientele to whom announcements, newsletters, and information are routinely sent.

The symposia and publications of the Society exhibit many different and personal viewpoints. Indeed, dissent is encouraged as a means of filtering, testing, and refining the professional skills of its members and the soundness of their work. Consequently these products, of which the present volume is a part, represent the views of their authors, and not of the Society or the United States Government.

Dedication

TO THE MEMORY OF

Commander Richard H. Barchi, US Navy: MORS Director, analyst, and gentleman.

Foreword

There comes a time in every profession when it is appropriate to record the core of wisdom generated by and shared among the practitioners and theorists of their craft. The origin of this monograph was the conviction shared by four past Presidents of the Military Operations Research Society that such a time had come. Military operations research has been *practiced* for forty years; arguably for a great deal longer. The military operations research *profession* has existed formally since the establishment of degree granting programs in operations analysis nearly concurrently at both the US Naval Postgraduate School and the Case Institute of Technology in 1951. The profession of operations research in fact antedates the popularity of the term *model* itself, though not the use of models.

During that forty years we have learned a great deal about modeling, expanding on the seminal theories of such pioneers as Bernard Koopman, P.M.S. Blackett, Philip Morse, George Kimball, and Glenn Kent. But mostly we have taught ourselves, in the school of hard knocks, about both uses and misuses of models.

The generations that flourished in those first forty years are yielding to a new generation, with its own set of attitudes and tools. It is a computer generation, proficient in hand-held calculators and desk top computers, born and raised on Avalon Hill war games and Atari simulations. It is a generation that believes large scale integration is old hat and takes the future of artificial intelligence for granted. The very power of modern modeling tools and the attitudes, half-unconscious, they generate impel a review of the strengths and limitations of models of every description for every application in defense analysis.

The centerpiece of this monograph is the Overview. It was drafted by Wayne Hughes to describe military models across all of their dimensions. It was then subjected, three times, to review and revisions by a committee I appointed as then-President of MORS. The members of this committee were chosen for their breadth of experience, their diversity of viewpoint, and not least for the respect they have earned in the profession of military operations research. As will be seen, for once only a committee could have carried out a task of such dimension.

FOREWORD

The aim of the Overview is to discuss *orthodox* military modeling. The committee endeavored to collate the collective wisdom and experience of the profession, debating not their own predispositions, but what they ascribed to be the beliefs held by the military operations research community as a whole. Doubtless they failed in some particulars. Still I would emphasize that they found a surprisingly large set of commonly held lore, insights, and judgments. Moreover there may be underway a further convergence of views. This monograph may well serve to hasten that process.

Following the Overview are fifteen chapters written by fifteen expert practitioners. Taken together these chapters elaborate on and exemplify the general conclusions in the Overview. These fifteen chapters are intensely personal; each writer draws from his own experience, aiming for depth and richness of illustration, without necessarily attempting to cover all aspects of his subject. For example, both J.J. Martin and Thomas A. Brown review a considerable amount of history; others do not. John Friel, Wilbur Payne, and Robert Hallex not only describe the characteristics of good air, land, and sea battle models, but each does so with a different slant, drawing on his especial operational, conceptual, and technical strengths.

By way of contrast with the Overview, some chapter authors go beyond what is currently accepted practice and project what in their fields is likely to be desirable in the future. Martin, Payne, Stephen Drezner, and Richard Hillestad describe from their own perspective what should be done to improve modeling of nuclear warfare, land battle, and logistics. But these changes are all consistent with the principles and good practices in the Overview.

One of the early substantive focus points of this monograph was to put in better perspective what is variously called validation and verification. I believe the monograph achieves this aim. The Overview, with its concept of "corroboration" as the more accurate term, Clayton Thomas' Chapter 13, and Wilbur Payne's Chapter 14, together serve as a needed reference point for future discussions. The monograph is not the last word, but it is certainly a suitable first word by MORS on the subject of verification and validation.

FOREWORD

In the course of preparing this manuscript it became evident that other expertise would have to be tapped to write the remaining chapters. The other authors became more than chapter contributors; many became collaborators in the entire effort. Clayton J. Thomas in particular not only contributed two chapters but was of inestimable assistance in ways beyond description.

Early in its production the Overview draft was widely distributed to attendees of the Military Operations Research Symposium. They were invited to comment in writing or in person at the 48th MORS. Indeed, the two chapters by Alfred Lieberman of the Arms Control and Disarmament Agency and Stephen Leibholz, President of Analytics, Inc., stemmed from their criticisms of shortcomings, which they were then invited to rectify.

****

It is my pleasure to acknowledge the many contributors to this work. It is the brainchild of Charles E. Woods, then-President of the Military Operations Research Society and my predecessor. It was Woods' inspired choice of Captain Wayne Hughes, Jr., USN (Retired), of the Naval Postgraduate School, to lead this work. He marshalled the support of our four MORS sponsors: at the time, Major General Robert A. Rosenberg, USAF, Assistant Chief of Staff for Air Force Studies and Analysis; Walter W. Hollis, Deputy Under Secretary (Operations Research) of the Army; Rear Admiral John A. Baldwin, USN, Director of Navy Systems Analysis; and Rear Admiral Kleber S. Masterson, Jr., USN, Director of the JCS Studies Analysis and Gaming Agency. They and their representatives have been at once the committee's faithful champions and most astute critics.

As noted above, the review committee, chaired by Wayne Hughes, was comprised of Dr. James J. Martin, Senior Vice-President of Science Applications, Inc.; Colonel John Friel, USAF, then of AF/SA; John A. Englund, President of Anser, Inc. and past President of MORS; Dr. Bernard D. Rostker, Director of the Systems Management Division, Systems Research and Applications Corp.; Dr. Marion R. Bryson, Director of the Army Combat Developments Experimentation Command and past President of MORS; and Dr. Wilbur B. Payne, Director of the Army TRADOC Operations Research Activity. From the outset this group volunteered their time not only to meet and hammer out the substance of the Overview, but to write a chapter each to expand on their corporate conclusions.

Thanks also are due to Thomas Anger of the Center for Naval Analyses for his constructive editorial review; to Edward Napier, MORS Executive Secretary, who contributed both to the substance of the monograph and its publication; to Ms. Ellen Saunders of the Naval Postgraduate School for her massive administrative assistance; and to the staff of Analytics, Inc., who helped with publication in many ways.

Special appreciation belongs to a final set of reviewers: Major General Jasper A. Welch, USAF (Retired), long-standing MORS supporter; Charles Woods of CNA, who thereby helped at both conception and birth; John K. Walker, Jr., of Rand and also past President of MORS; and the sponsors' representatives to MORS -- Clayton Thomas, whose direct contribution I have already lauded; Dick M. Lester, who was one of our most constructive critics; and Dr. Francis E. Shoup and LTC Anthony J. Quattromani, USA, both of whose faith in the ultimate quality of this work never wavered.

All of the collaborators gave their time without compensation. Therefore it seems fitting to close with thanks to all of the parent organizations -- and I have endeavored to acknowledge many of them -- and to the families of the participants for suffering the days away at meetings and the nights of writing that made this monograph possible.

Washington, DC
April 1984

Amoretta M. Hoeber

Introduction

From the outset, our purpose in drawing together a committee on military modeling has been to record the collective wisdom of the military operations research community, with these aims:

- to describe the attributes of well-conceived military models (and their limitations);

- to show how models can contribute to the decision process;

- and to remind the reader how models can be misused and oversold.

Our notional reader was, and still is, the military or civilian professional who has a grasp of military operations and the defense decision process, but only a superficial acquaintance with the models that inform the process. He may be a decision maker who must draw from the modelers' products or a fledgling study participant or study monitor. This "model" of our reader is still before us.

Along the way we added another set of notional readers -- the military operations research community itself. Why? Because we can all profit by periodically taking a broad look at the applications of military models as an antidote to our narrow view of what a model is and how it should be used.

For many an analyst, "model" flashes a vision of a computer simulation of a battlefield. For him the art of model building is the construction, maintenance, and improvement of complicated machine simulations that may be put to use for a variety of applications -- what we will refer to later as standing models. You will find many such models listed in the two-inch-thick *Catalog of Wargaming and Military Simulation Models* issued by the Joint Chiefs' Studies Analysis and Gaming Agency (SAGA).

For another analyst typical of many in the studies community, a model is constructed *ad hoc* to aid an analysis for a client who faces a decision. This analyst tailors his measure of effectiveness, input variables, and parameters according to the specific issue at hand.

INTRODUCTION

For decision makers at all levels who understand and appreciate the utility of models, the image may be a fairly simple set of mathematical relationships that "get right to the important variables," all of which will fit easily on the back of the proverbial envelope.

For the second- and third-echelon administrator who must carry out a high-level decision, "model" probably conjures up linear or dynamic programming, network theory, an inventory model, or a cost model for developing the detailed logistic and management support plans that must accompany major operational or procurement decisions.

For a research analyst at NASA, a model may be a differential maneuvering simulator in which a military pilot "flies" a new fighter in mock combat, the motion of which is governed by a set of equations representing his aircraft's aerodynamic characteristics.

For a researcher at the Naval Weapons Station, China Lake, a model may simulate a radar, radar jammer, or radar homing missile.

Thus we have much to say both to the novitiate in the world of models and their applications, and to the professional who has not thought lately about the *other* uses to which many types of models are being applied.

A difficult problem we constantly faced was spilling over the border line between the model and the analysis it served. It is impossible to discuss the proper design of a model's processes, its inputs and its outputs, without entering into matters of good analytical practices. Where the Overview discusses analysis and studies it is for the purpose of discussing models; the subject of good modeling is broad enough. Some chapter authors, however, deliberately cross the model-analysis interface. For example, Thomas (Chapter 1) observes that it is impossible to talk about wartime models except as they are interwoven with the underlying analytical problems, so *ad hoc*, vital, and immediate are these problems. Others, especially Lieberman (Chapter 10), point out the phenomenon of broad agreement on the model, but vigorous disagreement on the analytical application as to choice of data, scenario, and measures of effectiveness.

INTRODUCTION

The 40-year history of military OR is a compression of the history of science itself:

- a discovery of the power of disciplined thought to solve practical problems;

- a probing of the underlying structure of phenomena and the construction of theories and models that expanded our understanding of these phenomena;

- an explosion of optimism culminating in an "Age of Reason" in which it was said that all the universe was going to come under the sway of logical thought, and the scientific method was the special path to knowledge and verity;

- and an age of disillusionment in which idealists turned cynical, patrons turned scornful, and ultimate truth continued to recede beyond grasp.

The operations research community ended the 1970s having largely corrected the excesses of the 1960s. But it was not yet ready to shed the sackcloth of self-criticism and get on with its business without apology and without exaggerated claims. A positive theme of this monograph is the pervasive utility of models for military inquiry and decision making.

Among the many quotations that express the relationship between intellect and willpower in war -- the roles of head and heart -- one from Clausewitz seems to capture the essential role of military analysis and *particularly* of modeling:

> If theory investigates the subjects which constitute War; if it separates more distinctly that which at first sight seems amalgamated; if it explains fully the properties of the means; if it shows their probable effects; if it makes evident the nature of objects; if it brings to bear all over the field of War the light of essentially critical investigation -- then it has fulfilled the chief duties of its province.

But understanding the role that intellect plays in planning and winning war is only the beginning of wisdom. The most important thing we would have the reader take away from this work is a better understanding of the importance of experience, knowledge of subject matter, knowledge of technique, and -- in the very best work -- the genius of model creativity.

Wayne P. Hughes, Jr.

CAPTAIN WAYNE P. HUGHES, JR., US NAVY (RETIRED)

Commissioned from the US Naval Academy in 1952, Hughes commanded the minesweeper HUMMINGBIRD in 1957-58, the destroyer MORTON in Vietnam operations in 1968-69, and was Chief of Naval Education and Training Support from 1975-77. In operations research, he served as the first antisubmarine warfare analyst in OPNAV's new Systems Analysis Division in 1966-68; senior tactical analyst at COMFIRSTFLT in 1969-70; Deputy Chief of Operations Research on the SACLANT staff in 1970-72; and Assistant Chief of Staff for Analysis for COMASWFORLANT and (when the staffs were combined) COMSECONDFLT in 1972-73. He returned to OPNAV as the Deputy Director of Systems Analysis in 1973-75. From 1977 to 1979 he was Executive Assistant to the Under Secretary of the Navy. Since 1979 he has been on the faculty of the Naval Postgraduate School. His publications include two prize essays in the US Naval Institute's annual competition. He has served twice on the MORS Board of Directors. He holds a Master's Degree in Operations Research from the Naval Postgraduate School, where he was elected to the Sigma Xi.

Military Modeling

Table of Contents

The Military Operations Research Society iii

Dedication iv

Foreword .... Amoretta M. Hoeber v

Introduction .... Wayne P. Hughes ix

Overview

A. The Nature of Models and Modeling 1
B. The Role of Models 12
C. Selecting a Model's Properties 18
D. The Attributes of Models for Seven Applications 22
E. General Principles and Practices 33

CHAPTERS

1. Models and Wartime Combat Operations Research 55
Clayton J. Thomas

2. Modeling Nuclear Warfare 95
J.J. Martin

3. Air Battle Models 127
John Friel

4. Ground Battle Models 147
Wilbur B. Payne

Table of Contents

CHAPTERS

5. Sea Battle Models 165
Robert A. Hallex

6. Human Resource Models: An Overview 187
Bernard D. Rostker

7. Logistics Models: Evolution and Future Trends 203
Stephen M. Drezner & Richard J. Hillestad

8. Weapon Procurement Modeling 225
John A. Englund

9. The Role of Modeling in Force Sizing 235
Thomas A. Brown

10. Models in National Policy Analysis 247
Alfred Lieberman

11. Experiments 269
Marion R. Bryson

12. The Misappropriation of Models 281
John A. Battilega & Judith K. Grange

13. Verification Revisited -- 1983 293
Clayton J. Thomas

14. Validity of Models as a Basis for Military Planning 315
Wilbur B. Payne

15. Twenty Questions 327
Stephen W. Leibholz

Index 351

Figures

Figure 1	Bivariate Normal Surface	9
Figure 2	The Model Spectrum and Characteristic Trends	10
Figure 3	Mathematical Approximation by Two Terms of a Series Expansion	45
Figure 4	Approximation: Effect of Adding a Third Term	45
Figure 5	Estimating a Non-Linear Function with Changed Data $\Delta Y/\Delta X$	47
Figure 6	Classification of Army Models	148
Figure 7	The Hierarchy of Combined Arms Combat Models	154
Figure 8	Family of Land Battle Models	158
Figure 9	The Logistics Model	205
Figure 10	Strategic Counterforce Exchange (Generated Alert)	258

Overview

A. WHAT MODELS ARE -- THE NATURE OF MODELS AND MODELING

> "Today it is not the word 'truth' but the word 'model' that continually decorates the pages of scientific journals."
>
> -- Owen Gingerich†

A model is a simplified representation of the entity it imitates or simulates. DEFINITION

Speaking generally, a model may be a miniature of an airplane or a tank (an iconic model). It may be a map: an artificial representation of the earth in which heights are given by contour lines or vegetation by colors (an analogue model). It may be a symbolic representation of the object: a word description or a numerical description (a symbolic model). The language of models is all around us because models are ubiquitous. TYPES OF MODELS

In operations research the goodness of a model is judged by how well it achieves its purpose. This is no casually selected criterion. Operations research is in the main an applied science more akin to engineering and economics than to the physical sciences. In the physical

---

†Owen Gingerich, "The Galileo Affair," Scientific American, August 1982, page 137. Gingerich is illustrating the modern viewpoint that what passes for truth in science is only the most probable, always provisional, and never final. Thus the new fashionability of models as hypotheses. In particular he ascribes to Galileo a process of reasoning that "was, in an embryonic state, what is now called the hypothetico-deductive method: the testing of a hypothetical model, which attains ever more convincing likelihood as it passes each test successfully."

sciences a model is usually judged on the basis of, first, how accurately it portrays the phenomenon it represents, and second, its range of applicability. Many times our standard of goodness is dissatisfyingly vague: who would not prefer an image of universality such as, say, the laws of motion, which are very powerful models indeed. We will, by contrast, have unkind things to say about "universal" models within operations research which try to serve too many purposes. We will be devoting this Overview to a better understanding of models and ways to judge their goodness.

MODEL GOODNESS

A model is a shorthand form of communication. Some of the most powerful models not only describe an object, like an artillery battery, but also describe a process. Thus $\frac{dN}{dt} = -\alpha MN$ is, with suitable definition of its symbols, a mathematical statement of the process "area fire." Quantities used in such models communicate precisely and unambiguously. Seven is more precise than "a few." $P_d = .8$ conveys more specific meaning than "a high probability of detection."† The rigor of numbers and mathematical models sharpens the debate. "A few" may be indisputable, whereas "seven" may be taken as an estimate, a mean value, or some other shorthand for much more than the mere number. But "a few," for all of its comfortable security, seldom leads anywhere.

MATHEMATICAL MODELS COMMUNICATE WITH PRECISION

Statesmen, sociologists, and psychologists deal in shades of meaning -- subtleties of expression -- that the mathematics of science cannot match. These professionals also have their (symbolic) models. We do not understand the written word until we are prepared to wrestle with the analogical nature of language. "Social democracy" is shorthand for a complex -- and imprecise -- reality. "Bureaucrat," "politician," and "statesmen" are descriptors that paint mental pictures rich in meaning. They are models: purposeful abstractions from a more complex reality. Everyone uses models, even when he recognizes their shorthand nature. But word models are not as readily processed as the quantitative models of operations

WORD MODELS ALSO COMMUNICATE ANALOGICALLY

PROCESSING POWER OF OPERATIONS RESEARCH MODELS

†A numerical range band is arguably more accurate. It could derive from a standard deviation, a set of parametric variations, a confidence interval, or just a hunch. But a single value is more easily processible, and as we shall see, that is important.

research. With a numerical model, one can experiment, calibrate, and explore quantitative relationships.†

SIMULATIONS AS EXPERIMENTAL VEHICLES

A simulation, which may be iconic, analogous, or symbolic, does more than represent the phenomenon: a simulation imitates it. With a battle simulation one combines men, tanks, terrain, movement, and firepower to study their interrelationships and composite effects. The simulation is an abstraction, but it is a powerful abstraction when wisely employed. In effect, one conducts an experiment with the simulation model instead of with the phenomenon itself: he does vicarious experimentation. And simulations, including digital computer simulations, can be manipulated in remarkable and informative ways.

TWO OUTSTANDING INTRODUCTIONS

The reader who is not already acquainted with the nature and application of models in the more general sense will be rewarded by reading further on the subject. Chapter 10, entitled "Models," in E.S. Quade's *Analysis for Public Decisions* (1975) is a cogent overview of models, including concepts of non-numerical models for decision making. Another fine introduction is E.A. Bender's little book, *An Introduction to Mathematical Modeling* (1978). It leads its reader into the large world of symbolic mathematical models in an engaging, comprehensible way.

Military Model Taxonomies

MILITARY MODEL DEFINITION

For our purposes we define a military model as an abstraction of reality, the elements of which are chosen for (a) an investigative purpose or (b) a resource management purpose; in other words, an abstraction to assist in making decisions.

MANY TAXONOMIES EXIST

There is no one accepted way to categorize military models. Several systems that are similar in structure have been advanced by Ackoff (1962), Low (1977), Battilega and Grange (1978), Brewer and Shubik (1979), Taylor (1980), F. Hoeber (1981), and others. We will draw on these

---

†*One of the intriguing and propitious areas of modern model research and application is in processing "word models." The foundation was layed in that discipline within philosophy called logic. Boolean algebra and set theory were milestones in establishing linkages between words, ideas, symbols, and numbers. Presently work in fuzzy sets and some aspects of what is broadly termed artificial intelligence are being applied profitably in modern computer simulations.*

writers, but establish our own categories, none of which we regard as fixed or final.

1. Models by Technique or Level of Abstraction. Models with which we deal range in a continuum encompassed by:

FOUR TECHNIQUES

(1) An analytical representation

(2) A computer simulation

(3) An assemblage of war gaming people and tools

(4) A field experiment†

2. Models by Application or Purpose. We define seven areas of model application here, discuss them in Section D, and elaborate on them in Chapters 1 through 10. These applications are for:

CATEGORIZATION BY APPLICATION

- Battle Planning: the preparation for wartime operations, based on friendly and enemy orders of battle, existing strategic or tactical environment, and specific missions or objectives.

- Wartime Operations: the conduct of the war. Distinguished from battle planning by available, current, combat data and known, immediate military objectives.

- Weapon Procurement: the selection from among competing weapon systems or characteristics for procurement in the near future.

- Force Sizing: deciding how many weapons of which types to (1) operate, (2) support, and (3) procure for the future, either in the defense establishment as a whole, or in a major component such as the Army or the nuclear weapons arsenal.

- Human Resource Planning: the design and operation of manpower, personnel, training, and assignment systems.

- Logistic Planning: the design and operation of all manner of military logistic support.

---

†*See Note 1 at the end of the Overview for a discussion of these terms.*

- National Policy Analysis: supramilitary actions that influence or are influenced by military considerations such as arms treaties, or subsidies of commercial transportation systems.

Two other prominent categories of models which are imbedded in two or more of the above categories, but which are often treated by themselves and have an extensive literature of their own, are:

- Command, control, communications, and intelligence models, sometimes called information warfare models.

- Cost models.

3. Models by Scope or Scale. We will refer to three levels of model aggregation for battle modeling:

- Micro or single-unit engagement models (such as submarine versus submarine, fighter versus bomber, or attacking battalion versus defending company).

- Medial or multiple-unit engagement models (such as a naval battle group versus combined air and submarine attackers, or an Army divisional assault on a defending brigade, or a formation of air intercepters versus attacking fighter-bombers).

MICRO, MEDIAL, AND MACRO MODELS

- Macro or theater-level model (such as the defense of shipping in the Atlantic over an extended campaign or air-ground combat across the Central Front in Europe).

4. Ad Hoc and Standing Models. Occasionally we distinguish and contrast the characteristics and benefits of two model types, which we refer to as:

- *Ad hoc* models that are built with a specific decision in mind, and

- *Standing* models that are maintained, operated, and improved on a continuing basis and which usually serve decision making less directly and less obviously than *ad hoc* models.

5. Models to Describe, Prescribe, and Predict. A *descriptive* model reproduces essential processes of the phenomenon it models. A combat simulation model is descriptive: it imitates parts of the battle. Sometimes an insightful description of the process being modeled

MODELS TO DESCRIBE

alone is adequate to imply a course of action. (See Thomas, Chapter 1, page 82 .) An inventory model which records and displays the location and numbers of spare parts and personnel is not merely descriptive, but highly utilitarian in the right circumstances.

...TO PRE-SCRIBE

A *prescriptive* model specifies a course of action. Linear programming, dynamic programming, game theory, and decision theory are methodologies that solve problems in ways that tell one what to do. For example, game theory solutions specify the optimal strategy to follow and the payoff that is guaranteed under the strategy. If the inventory model mentioned above which records and displays stock levels also specifies when to reorder stock by some rational decision criterion such as a variation of the Economic Order Quantity (EOQ) model, then the model becomes prescriptive.

...AND TO PREDICT

It is doubtful that there are *predictive* models which are entirely distinctive from descriptive or prescriptive models. However, we want models with predictive power. When we are satisfied that a model describes an existing situation or phenomenon adequately, then we want to apply it to process input data and arrive at results in other situations.

"ROBUSTNESS" AND ITS CRUCIAL IMPORTANCE

The wider the range of conditions over which the model has been demonstrated to predict experimental results, the more "robust" the model is said to be, and the greater its utility. Many arguments associated with model verity are really arguments over model robustness. Military models that describe some circumstances very well may fail when applied to dissimilar circumstances. Scientific models like the laws of thermodynamics and motion are imposing for their demonstrated breadth of applicability and (depending on the engineering application) predictive power and accuracy. Often, probably most often, operations analysts must adapt models with less prestigious pedigrees. In Section E, page 36 , we will caution against the misappropriation of model *results* for circumstances to which they do not apply. But here we are more concerned with misuse of the models themselves, including an awareness that the sufficiency of the model usually depends on the range of applications and inputs in those applications.†

---

†See Note 2 at the end of the Overview for a more mathematical approach to the foregoing. In addition Chapters 12 and 14 illustrate the misapplication of operations research models.

In practice models seldom can be neatly characterized as descriptive, prescriptive, and predictive. Descriptive models are used prescriptively to search out superior solutions by educated trial and error: we say by "brute force." Prescriptive models are used for insight as well as decision. Some loss of fidelity is usually sacrificed to incorporate their optimizing properties. Linear programming models, for example, require linearization of relationships that are typically curves over some range of inputs. Therefore prescriptive models are wisely used in conjunction with more refined descriptive models. Whether the modeler accepts the solution of a prescriptive model literally as the basis for a recommendation, or whether his client adopts the course of action recommended, depends on an assessment of the adequacy of the model and the range of conditions over which the model is thought to hold.

GREATER LIMITATIONS OF PRESCRIPTIVE MODELS

6. By Properties: Transparency, Reproducibility, etc. A list of fourteen model properties was prepared by the Army Models Review Committee and used for evaluating Army models.† They were:

(1) Consistency

(2) Enrichment potential

(3) Experimental validity

(4) Military realism

(5) Physical reasonableness

(6) Visibility to the analyst

(7) Credibility

(8) Flexibility

(9) Interface potential

(10) Resources required

(11) Responsiveness

(12) Sensitivity of the model

---

†*Taken from Brewer and Shubik (1979), page 329.*

(13) Technical user capability

(14) Visibility to the user

TRANSPARENCY

"Visibility" to the analyst or to the user is often referred to as model "transparency," or the understandability of the model and its results. A field experiment is transparent to the Army general ordering the exercise because he understands how Army forces operate in the field and he has confidence in the results -- sometimes too much confidence as Bryson discusses in Chapter 11, "Experiments." Analysts usually think of analytical models as *transparent* because they have worked with and understand the relationships in the mathematical formulas. But if the analyst is applying a concept as "simple" to him as a bivariate normal probability density function,

$$p(x,y) = \frac{1}{2\pi(1-\rho^2)^{1/2}\sigma_x\sigma_y} \exp\left\{\frac{-1}{2(1-\rho^2)}\left[\left(\frac{x-\mu_x}{\sigma_x}\right)^2 - 2\rho\left(\frac{x-\mu_x}{\sigma_x}\right)\left(\frac{y-\mu_y}{\sigma_y}\right) + \left(\frac{y-\mu_y}{\sigma_y}\right)^2\right]\right\}$$

then he had better be prepared to augment the formula with a picture like Figure 1 and explain how his hillock peaks, flattens, stretches, and shifts on the plane as the five parameters change in the mathematical model.

OPACITY

Of course complicated computer simulations are notorious for the property opposite of transparency: *opacity* to the client and often even to the analyst. But the simulation is opaque at least in part because it will not have been scrutinized and tested by as many authorities as the bivariate normal. Here let us be content to say that opacity is a deadly deficiency and transparency comes from the amount of polish devoted to the model: it is a product of hard work, usually unappreciated.

REALISM AND RELEVANCE

An important distinction is between *realism* and *relevance*. As Brewer and Shubik say in *The War Game* (1979), military realism and physical reasonableness are certainly desirable qualities, but realism and relevance are not equivalent. By their very nature models are unrealistic, and it is important to build and use, as they put it, "'correctly' unrealistic models."

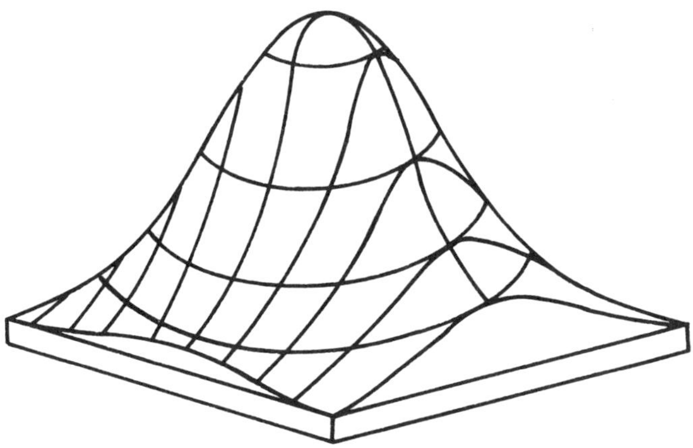

Figure 1. Bivariate Normal Surface

An axiom of systems analysis is that models must be *reproducible*, which means that another user should be able to get the same results with the same inputs. One of the weaknesses of a large-scale computer simulation model is the tremendous effort it would take a skeptic to reproduce it. Since the model is written in some computer language, it is reproducible in principle, but a more useful practice is to compare results of models thought to be similar in structure using similar inputs. If the results are far apart, there is a problem. See Lieberman's discussion in Chapter 10, "Acceptance of Models" (pages 251-253), for one example of this useful approach to "reproducibility." For war games and field experiments "reproducibility" takes an even more pragmatic bent. We neither discard the results of games and experiments because there are outcome variations caused by the personalities of the participants, nor discard the objective of reproducibility by some reasonable standard. Bryson discusses standards of reproducibility as regards field experiments in Chapter 11. As war gaming increases in popularity for investigative pur-

REPRODUCIBILITY . . .

A GOAL OFTEN UNREALISTIC

purposes, techniques to identify and isolate player effects will be an important object of research.

A valuable way of displaying typical trends of some model characteristic is illustrated in Figure 2, adapted from Lawrence Low (1977) and James Taylor (1980). Operational realism, the degree of abstraction, outcome reproducibility, and convenience and flexibility are related to the different techniques of modeling as shown.

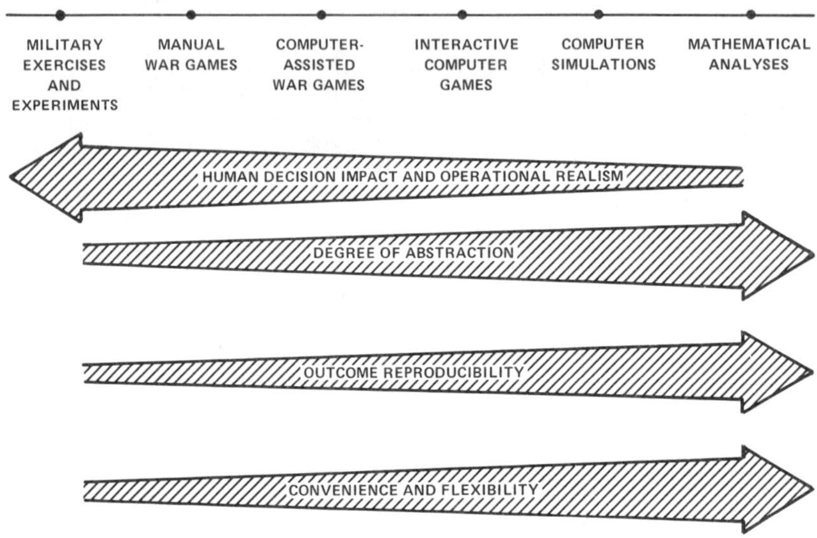

Figure 2. The Model Spectrum and Characteristic Trends

"SCIENTIFIC" IMPLIES OBJECTIVITY

7. Scientific and Sensible Models. Military operations research is regarded as a science for two main reasons. The first is objectivity. Because we serve clients who make life-and-death decisions involving large amounts of power, prestige, and money, sharply polarized debates often arise between analysts who are serving opposing sides as, in effect, expert witnesses. After a particularly acrimonious dispute in 1969, the Operations Research Society of America, or ORSA, convened an Ad Hoc Committee on Professional Standards. The committee report (1971)† reaffirmed that the standard of professional practice was to be open, objective, and explicit analysis,

---

†*Contained in The Journal of the Operations Research Society of America, Volume 19, Number 5, September 1971.*

conforming to the scientific spirit of the pursuit of knowledge. The committee went on to offer guidelines for the operations analyst in an adversary process, which acknowledged the inevitability of personal conviction and partisanship, but eschewed advocacy -- the presentation of facts supporting only one side of the case -- as unacceptable practice.

"SCIENTIFIC" ALSO IMPLIES VERIFICATION

The second claim to status as a science is the standard of scientific proof, which demands that theories be open to examination, subject to peer review, and "verifiable" by experimentation ("falsifiable" is the scientifically correct term). The requirement for verifiability must be tempered in practice because some models cannot be tested under realistic conditions. Indeed, in the case of nuclear warfare, a model may have been constructed in the first place because *no* test was possible.

CORROBORATION

To enhance the model's robustness, the wider the range of test conditions the better. The degree and extent of correspondence between predictions and experimental results determines how well, and over what range of inputs, the model is thought to be verified, or corroborated.† Most models can be tested only partially, and in limited conditions. Obviously, modelers should strive to find ways, direct or indirect, to corroborate their work, and should be careful that the extent of testing is made clear.

Nevertheless, models often must be relied on for their predictive power -- as in the early stage of weapon procurement planning -- when they cannot be tested. Then, the model and all that flows from it are merely hypotheses, unsubstantiated and unverifiable.

SENSIBLE MODELS

It is best to think of such a model not as scientific but as a logical or "sensible" construct -- open, explicit, and reproducible. In "sensible" modeling, the parameters and variables are set down; the processes (equations or simulation flow diagrams) are open to inspection; and the same inputs will reproduce the same outputs. The

---

†For a discussion of the terms verification, validation, and corroboration, see Note 3 at the end of the Overview. For more thorough discussions see Chapter 13, "Verification Revisited -- 1983," by Thomas and Chapter 14, "Validity of Models for Military Planning," by Payne.

structure of the model and the choice of inputs become the focus of debate. The results simply follow. Once a sensible model is accepted as no more than it is, then it can be evaluated on the proper basis: how well it helps to structure debate over a decision at hand by virtue of its transparency, flexibility, and reproducibility, and above all, by how well it incorporates the essentials of the process it represents.

THEIR STANDARD OF ACCEPTABILITY

Even though sensible models are transscientific (beyond verification), they are nevertheless subject to rigorous scrutiny, which is the essence of science. Everyone who has devised a model that was used in arriving at an important decision knows the vigorous debate that occurs over the suitability of a model and its inputs. Dorothy L. Sayers once wrote about characters in fiction:

> No considerations of false reverence should prevent us from subjecting the incarnations of creators to the severest tests of examination... they should be pulled about and subjected to the most searching kind of inquiry. If the structure is truly knit, it will stand any strain, and prove its truth by toughness.

It is through such inquiry that sensible models must be tested and found to ring true or false.

B. THE ROLES OF MODELS -- HOW MODELS ARE USED

"Models are for thinking with."

-- Sir M.G. Kendall

Military Models Are Decision-Oriented

This Overview is about *modeling*; insofar as possible we avoid discussions of decision making, studies, and analyses. But because models are servants designed to contribute either immediately or eventually to problem solving, we cannot talk about modeling in isolation. We will use these terms:

- *Analysis* exercises a set of assumptions and inputs in a model to give results and conclusions.

ANALYSES, STUDIES, AND CLIENTS

- *A study* is a comprehensive examination of a problem, including an analysis. A study should comment on the incommensurables the analysis did not treat. Typically a study carries recommendations.

• *The client* is the user of the analysis or study: the sponsor, the initiator, or any decision maker who draws from the effort.

THE BASIC PROBLEM-SOLVING STRUCTURE

The client has some goal or objective which, if attaining it is a problem worthy of study, involves alternatives -- possible ways to attain the goals. In order to understand the consequences of each alternative he needs an analysis. He will need some decision criteria and measures of effectiveness to judge the attainment of the criteria and the costs (penalties) paid to attain them.

THERE IS ALWAYS AN IMPLICIT MODEL

The design of a model -- perhaps several models -- is in practice almost indispensable in conducting this orderly, structured process of decision making. One should not infer that every decision generates a *new* model. After 40 years of practice, few models are designed anew: mostly they are adaptations or, for persistent problems, products of evolution. Still, every logical decision maker has some sort of mental model whenever he makes an orderly decision. The model -- by definition -- does not capture every aspect of the client's problem. The analyst tells the client what was and was not modeled and the reasons. If the analysis is incorporated in a study with recommended courses of action, then the analysts should have given weight to factors that were not modeled and qualitative aspects of the problem that could not be analyzed. Since no team of analysts can substitute completely for their client -- the decision maker -- study recommendations are always provisional.

ON CLIENT PROXIMITY

The closer a decision maker comes to conducting the study himself (as for instance, by manipulating simple mathematical relationships) or to actively participating in the selection of decision criteria, scenarios, assumptions, and model properties, then the more likely that the study recommendations will be implemented literally. Since this is obvious, it is astonishing the amount of modeling, analysis, and studying that goes on for clients who believe it is merely the final product that they are purchasing.

The biggest payoff comes from active client participation throughout the process, in which case the client deserves a model that he can be made to understand, or, since this is not always possible, in which his confidence may be built. Models (and studies) provide the following benefits in proportion to the client's participation:

- help explore issues in an orderly way
- help structure and discipline the debate
- compare the virtues and vices of alternatives
- give insights and reveal new characteristics
- sometimes provide unexpected answers, which will probably seem obvious after the fact.

But models (and studies) do not:

- make the client's decision for him
- substitute for hard thinking
- answer unasked questions.

THE PROCESS AS THE PRODUCT

Like M. McLuhan's "The medium is the message," the process of structuring the problem, deciding on inputs and parameters, and exercising the model is usually more beneficial than the product.

The great art of modeling is to identify the primary relationships pertinent to the issue, isolate them, and study their effects.

Indirect Decisions

WITH MODELS, NOT ALL DECISIONS ARE IMMINENT

While it is right to think of military models as being decision-oriented, it is wrong always to key them to some imminent decision, much less to some high-level decision maker such as the Service Chief of Staff or the Senate Armed Services Committee. The utility of models is often more elusive and pervasive.

THE CUMULATIVE EFFECT

A highly placed and unfriendly defense official once asked the Navy's Director of Systems Analysis to list all the studies that had recently been used directly to make an important weapon procurement decision. After much effort, he found two. But he went on to say that the number of studies that had *influenced* decisions were manifold. Some of the studies of theater-level operations that simply provided a background of understanding of Navy readiness and employment concepts are cited by Hallex in Chapter 5. The results of models sometimes have long-term cumulative effects, modifying attitudes of both the analysts and their clients. There is almost always a gestation period before a new weapon or tactical concept takes hold. Model results influence attitudes that may be invisible even to a reluctant beneficiary.

Another indirect effect will be invisible to high-level decision makers in Washington to whom the only models which are relevant are those that assist them in their decisions. To quote from Wilbur Payne:

> There are a large number of specific quantitative decisions derived from modeling under circumstances in which I could not even locate the "clients" without effort. It is probable that in some real sense these decisions are derivative: they arise because of higher echelon policy or program decisions. Either individually or cumulatively these decisions are important, and many, if not most, employ models. The high-level decision makers should, from time to time, worry about the validity and objectivity of the process by which these implementing decisions are made.
>
> I do not suggest that the Secretary of Defense should worry about who is going to buy how many spare fan belts when he decides to buy more or fewer or different trucks. He may be satisfied to know that spare fan belts were costed into the alternatives presented to him. But when we build our models we should remember there is a fan belt buyer out there who is or ought to be using an inventory theory model. His job performance will improve if he uses better models, and we shouldn't wait until he knocks on our door as an official "client" before we help him to do so.

SUPPORTIVE DECISIONS

Models for Diagnosis

Models used diagnostically are largely invisible outside of the operations research community. Diagnostic work is chaff that never goes into study reports. Among models that serve the diagnostic purpose, the better known and understood tend to be complicated and detailed, typically computer simulations or semi-computerized war games.

DIAGNOSTIC MODELS

These are useful for finding driving parameters and aggregating or consolidating them. They help to discard extraneous parameters with confidence.

ABSTRACT DIAG-NOSTIC MODELS

The second category of diagnostic model is highly abstract. It is useful to explore relationships when the scenario is unknowable and the forces (or weapon alternatives, or new tactics) are partly conjectural. The analysis is done by tinkering with the model's parameters and inputs. Therefore such a model is simple and analytical enough to allow wide variation of a few inputs. For example, we use the sonar equation, in both active and passive forms, to model acoustic detection. We use abstract mathematical models of weapons effectiveness; the Army uses firepower indices; and others use indices for nuclear weapon stockpile comparison (which strictly speaking are indices rather than models of physical reality) such as Equivalent Megatonnage, Equivalent Weapons, or Countermilitary Potential. We use the cobweb model in economic theory, and we use war games to explore a new tactic in an ASW trainer with abstract water conditions at an unspecified location. All of the above can be embellished in an attempt to add realism, and most have been -- some successfully and others to questionable advantage. The most famous diagnostic model of all is Lanchester equations: coupled differential equations which, for all their limitations, are used again and again to explore some new aspect of attrition or force buildup, to investigate a new battle concept, or to explain some general principle.

WELL-KNOWN EXAM-PLES

A diagnostic model tends to be highly creative and *ad hoc*, easily† modified for specific variations and permutations. Often it will have been derived from insight that was gained from exercising detailed models with specific scenarios and forces, along with debate, experience, and plenty of hard thinking. An abstract model derived thereby is, as it were, a model of a model.

Models for Management

Models are also used in the Department of Defense for making management decisions. As is described by Rostker,

---

†We mean easy in the mathematical sense. It is a high art to grasp what mathematical modification represents the essence of the new tactic/weapon/manpower mix, or whatever is at issue, perhaps the highest in operations research.

Drezner, and Hillestad (Chapters 6 and 7), personnel, manpower, and logistic support are deployed, acquired, and otherwise processed with the help of complex standing models in continuous use.

HUMAN RESOURCE AND LOGISTIC APPLICATIONS

What is more interesting, there is a growing capacity for and advocacy of the appropriate use of models for battlefield "management." Decision aids are already evident for routine naval tactics: aircraft transit plans, sonobuoy search patterns, and ship ASW screening plans, for example. Advocates of the power of artificial intelligence go farther, and urge the development of battle planning models that advise, even direct, tactical actions under appropriate circumstances. Certainly in the past few years, new technology has achieved some practical successes in providing rapid, emotionless decision guidance in restricted fields which, by test, is on a par with guidance provided by human experts in tempo and quality. We also know that Russian scientists are exploring battlefield decision models in collaboration with Soviet military leadership. See, for example, Druzhinin and Kontorov (1972), Chuyev and Mikhaylov (1975), Gorshkov (1980), and especially Chapter 4, "Specific Methods for Automating and Optimizing Troop Control Processes," in *Automation of Troop Control* by Bondarenko and Volkov (1977).

GROWING INTEREST IN BATTLE MODELS THAT MANAGE

SOVIET INTEREST

Anyone who has had to develop decision rules in the world of computer simulations knows the difficulties modelers will face when they tackle "battle management" decisions of any complexity. Nevertheless models that assist in operational decisions -- for management itself -- are increasing in numbers, complexity, and importance.

CAUTIONS RE BATTLE MANAGEMENT MODELS

Use and Usefulness

At the outset we said that the goodness of a model had to be judged simply by how well it achieved its purpose. A model is useful if a better decision can be made with the information that it adds. Is it useful enough? We are reminded of one of MGEN Jasper Welch's quotable quotes: "For openers, one should do a cost benefit analysis of what's worth calculating." A model is produced and maintained at a cost. Not only is there a cost in dollars but in mental energy. All of us can recall times when after enormous effort was devoted to modeling a problem, in the end the main effect was to distract hard thought and discussion from the decision itself.

A STANDARD OF UTILITY

ON MODEL ENCHANTMENT

Worse, one can become so engrossed in the model that it takes on a life of its own. Like a superb actor who verily lives his part on stage, he comes almost to believe that his model world, that creation of his own energies, is the real world itself. Sometimes, we confess, a model is too beguiling. Even experienced analysts occasionally catch themselves likening combat models to physics models. We have to be reminded that military problems are hardly ever less complicated than engineering problems and usually are as intricate and involved as social science problems, rather more than less formidable than micro- and macro-economic analysis.

NOT TO FORGET: MODELS DEAL WITH COMPLEX SOCIAL INTERACTIONS

On a case-by-case, application-by-application basis, the analyst needs to give his client a fair estimate as to the range of uncertainty in the results. It is little wonder that military men who understand the limits of model prediction and the risks of military unpreparedness would like to add a margin for error when force levels and arms treaties are at issue.

C. SELECTING A MODEL'S PROPERTIES

"Pluralites non est ponenda sine necessitate." (Multiplicity ought not to be posited without necessity.)

-- The Principle of Occam's Razor†

The Great Guide: The Nature of the Decision

NATURE OF DECISION CLEARS UP UNCERTAINTY

Standing out like a beacon for the analyst is the question: how is the model intended to help a client with his decision? Much abstract argument over a model's suitability and applicability will be eliminated once its purpose is known -- and not before. The pedestrian process of pinning down the study objective, the decision criteria, the measures of effectiveness, the availability and quality of the data, and the scenario will go a long way toward determining the structure of the model.

PROXIMITY OF THE CLIENT

And what is the analyst's proximity to the decision maker? Will he often brief or talk with his client (who may be several people or a steering group)? Will the client be accessible and responsive or remote?

---

†*See Note 4 at the end of the Overview.*

What model will the client respond to? Will he put little or no trust in a model that is too complicated for him to fully understand? Contrariwise, will he distrust a model that is too symbolic, simple, and analytical?

TEMPERAMENT OF THE CLIENT

The Time to the Decision

Another crucial question is how long until the client must make his decision? The implications are many. First and most obviously, the analyst must respond within his deadline. There have been an amazing number of instances in which a client expected a model that could not possibly have been built or adapted and the inputs gathered in the time allotted. If the decision is imminent and there is no reservoir of data, expertise, and model work on which to draw, then one sees the backs of envelopes emerge.

DEADLINES DRIVE THE SCALE

Which raises the second point. It is rare when a problem is so new that the need for analytical assistance could not have been foreseen. As one of our number said, probably any senior analyst could generate a list of 50 decisions, including non-routine, non-repetitive decisions, that senior officials will have to make in the next three years and be right about most of them. We could generate another list of recent decisions that we suspect would have benefited from more analysis. The OR community has a responsibility to anticipate their clients' needs by maintaining its "industrial base" of models (or modeling methods), a body of expertise, and compatible data with which to treat recurring problems. At the same time the community of clients must foster and nurture this base and not assume that it will come into play automatically. The contributions of experts and the results of their continuing research are too often taken for granted by decision makers, or worse, their absence may go unrecognized along with the better decisions that might have been.

FORESIGHT ENHANCES RESULTS

ANALYST'S RESPONSIBILITY

There is a third matter. What does the modeler do when time is short, he is taken by surprise, and his instincts say his product will be defective? There comes a time to remain silent. Occasionally -- ah, the pain of it -- the back of the envelope should stay in the pocket.

A TIME TO DEMUR

But when the decision is imminent, the model will usually be simple: it may be profoundly simple or just simple-minded; the modeler will know which, and should tell his client.

The Goal of Simplicity

APPROPRIATE SIMPLICITY

Different in kind from model simplicity forced by time constraints is simplicity sought for better understanding. The perfect model would capture only the essential processes, and use only the essential parameters. Naturally the art of the analyst is to deduce what the right (minimum) properties are. As Albert Einstein once said: "Everything should be made as simple as possible -- but no simpler." A subsidiary problem is whether the analyst can convince his client that he knows which properties are important. Recalling Brewer and Shubik (1979), a client can confuse realism (a tank that looks and behaves like a tank) and relevance. The abstraction that captures only the necessary tank-like properties will be highly relevant, but will probably not be very realistic.

A DEVELOPMENT PROCESS

The discernment of what is relevant in a model comes from talent and experience. It is rare when all the essentials, but no extraneous phenomena, are settled on the first try. There is a development stage that good models go through. This is most evident of standing models with distinguished family trees, but it is also true of most of the best models that were developed *ad hoc* for use in a single study. The development process was embodied in the modeler's years of professional growth.

Conditions Warranting Model Detail

ADD DETAIL:

There are three circumstances that may justify the development of detailed models.

1. TO PROVIDE SUPPORT FOR *AD HOC* DECISIONS

When combat data are not available a detailed model helps provide a reservoir of raw pseudo-data and benchmarks (indices) for later *ad hoc* decision analyses, or as inputs to assist with policy decisions. Naval theater-level studies such as *Sea War 85* (see Hallex, "Sea Battle Models," Chapter 5) used a hierarchy of suitably detailed models to explore future scenarios. The models and the analyses were a long time in the making, the reviewing, and the assimilation. Neither a model nor an analysis can be thrown together when the policy issue arises: the "industrial base" must be in place.

2. TO PROBE MORE DEEPLY

A second justification for complexity is to examine in greater detail a contentious piece of a simple quantitative analysis. A good example is the detailed examination of an hypothesis in the early 1960s which held that the defense of Western Europe was feasible without resorting to nuclear weapons. The hypothesis stemmed from some simple quantitative, principally economic, analysis which had to be taken apart and examined bit by bit. Although the analysts who undertook this may have gone too far in trying to build very large, very detailed models, nevertheless there was a legitimate need for more detailed study of an issue with such important implications.

3. TO DEVELOP AND IMPROVE A STANDING CAPABILITY

The third circumstance arises when there is an abiding corporate or institutional client and when the decision process is drawn out or repetitive. In government, "decisions" tend to be strung out over many layers of review: made and unmade; deferred and accelerated; kneaded and massaged; done, undone, and redone. Wars comprise many battles with elements in common. Manpower and logistics decisions are often iterative and evolutionary. Such processes will foster the accretion of high-quality properties in their supporting models. When models emerge that are especially robust and utilitarian, then they ought to be perpetuated.

TRUSTWORTHY MODEL MANAGEMENT NEEDED

When the institutional client (we have in mind an organization like SAGA, AF/SA, or TRASANA) retains and maintains a standing model, then the model operators will learn many ramifications of the problem being modeled. It is important that the team that built the model be kept to exercise it, and that continuity be maintained in the team. The institutional client will benefit greatly to the extent that it is intimately involved, and often must become, in effect, the interpreter of the standing model's results for both higher levels and lateral levels of management. The influence of the model rests on the office's wisdom and persuasiveness, for no other office can hope to know a detailed model's strengths, weaknesses, and idiosyncrasies. Obviously there can be few such models in successful operation.

FEW SUCH MODELS POSSIBLE

Characteristics of Standing Models

THE NATURE OF "STANDING" MODELS

If every decision could wait for a model designed and tailored for it, then no standing model ought ever to be necessary. Yet all decisions are time constrained, and some of them will demand the support, at least indirectly, of models that take a long time to develop, refine, and understand. As noted a standing model is usually developed with some generic, repetitive set of problems in mind. The more common types of standing models are war games or computer simulations, sometimes man-machine interactive.

On the other hand, we caution that attempts to design all-purpose models to serve many problems and decisions have rarely succeeded. For one thing, the scope of the undertaking has been consistently underestimated. Since an all-purpose model's intended uses are universal, it must cover all aspects of every potential problem that it may be called upon to address. The bigger and more expensive it becomes, the more problems it must treat in order to justify its expense. In the 1960s especially, analysts attempted to build many such models by exploiting the capacity of high-speed computers. Instead of trying to determine in advance which were the important variables, their brute-force method treated all variables as important and assumed that the computer would overpower nature. As usual, nature won, overwhelming the computer and the analyst with her complexity. Even models at the micro level suffered from (1) the tedium of entering inputs; (2) the opacity of the processes; (3) the lack of data commensurate with the detail in the model; (4) the large number of parameters that had to be varied, with concomitant difficulties in determining the parameter combinations to which to give the most weight or credence; and (5) difficulties of corroboration, even though realism, physical fidelity, and validity were the motivations underlying them.

THE FOLLY OF ALL-PURPOSE MODELS

... AND WHY SO

In addition to many hazards associated with all-purpose models, standing models entail more mundane difficulties. A single agency must maintain configuration control over the entire hierarchy. Since the same model will often be maintained and operated in different locations by different commands, a system of change orders will be required. Then there are inevitable conflicts between rapid redesign to handle new demands and the need for stability and field personnel to keep model and documentation up-to-date. Computer hardware and technology are advancing so rapidly that hardware and software obsolescence become serious obstacles to model perpetuation. These truths hold whether one has in mind simulations, war games, or field exercises. Consistency is the enemy of adaptability.

REQUIREMENTS FOR STANDING MODEL SUCCESS

D. THE ATTRIBUTES OF MODELS FOR SEVEN APPLICATIONS

"What little theory we have about policy and systems analysis is mostly centered on models and model building."

-- E.S. Quade, *Analysis for Public Decisions* (1975)

In part the diversity of opinion over the value of different modeling techniques, the level of detail, and the pros and cons of analytical models, simulations, war games, and field exercises occurs because people have different applications in mind. Recall the seven applications of military modeling introduced in Section A:

(1) Battle Planning,

(2) Wartime Operations,

(3) Weapon Procurement,

(4) Force Sizing,

(5) Human Resource Planning,

(6) Logistic Planning, and

(7) National Policy Analysis

Battle Planning

Models for battle analysis are discussed first because they are probably the most common, because they are often adapted (both wisely and foolishly) for other purposes, and because they came first historically and in many ways intellectually.

EARLY MODELING

Battle models are associated with a long history of tactical analysis using quantitative methods that had flowered by at least the turn of this century. There are even earlier antecedents of tactical and strategic war gaming. Essentially all of the serious pre-World War II modeling was conducted by military men.

AIM TO IMPROVE COMBAT OPERATIONS

Battle models may be *micro*, *medial*, or *macro* in scale. Battle modeling aims to improve tactics or force composition by examining the effectiveness of new tactics, or different combinations of existing weapon systems. We will be careful here to discuss battle models only for the purpose of modifying tactics, deployments, or compositions. Because of the widespread employment of battle models for weapon selection and force planning as well, the chapters which follow take a broader view and address land, air, and sea battle and general nuclear warfare (including prehostilities crisis conflict), as these models are used for many investigative purposes.

DETAIL COMMON AND APPROPRIATE

A prominent characteristic of battle models is that detail is appropriate. Force compositions, geographical settings, weapon performance, and military objectives can be relatively well-defined because the modeler is dealing with a near-term problem. Fleet and field exercises are of this kind. Training purposes excepted, most war games are for battle planning and force evaluations. Even training war games and exercises may be used not merely to sharpen execution but to provide planning insights as well. Computer-assisted war games are especially useful. And as we have indicated, simple analytical models are also powerful for clarity of reasoning and for exposition.

RATES OF CHANGE

The most credible insights from a battle model will be contained in the rates of change. Attrition rate or movement rate information is far more credible than absolute casualty numbers after some length of time. Even when complex, dynamic decision rules are built into the simulation, they cannot hope to capture the two-sided process of action-reaction-counterreaction. The initial rate $(\frac{dX}{dt})$ will be more reliable than a later outcome, $X(t_1)$. However, the model should be run long enough to check for accelerations -- changes in the rates of change, $(\frac{d^2X}{dt^2})$ -- and when they appear every effort should be made to account for them. We emphasize the difficulty -- the near impossibility -- of modeling the real dynamics of battle beyond a few steps.†

ABSOLUTE RESULTS NOT NECESSARY

A good battle model yields trends, insights, and with due caution, order-of-magnitude indications of likely battle outcomes *under the given circumstances*. Happily, tactical or technological improvements are more easily deduced than battle outcomes. A simulation, or quite often a simple analytical model, can indicate the direction and relative improvement sufficiently to justify field tests and adoption of a new firepower mix, a new CAP station, or the like. Wartime effectiveness either before or after the change is a much more difficult evaluation.

OPTIMIZATION NOT NECESSARY, EITHER

Similarly one does not need to know that he has *optimized* in order to institute a change; an indication of significant improvement will suffice. One should not become overly enchanted with combat models that optimize.

---

†See Thomas's reference to P.M.S. Blackett in Chapter 1, page 66, and Note 2 at the end of the Overview.

They have their uses in exploratory calculations, but the simplifications and artificialities required in order to exercise an optimization model impel the supplementary use of other models of greater fidelity.

Wartime Operations

The emergence of operations analysis in World War II was the second key development in military modeling. The first operations research was wartime analysis done by scientists in collaboration with military men.

THE WARTIME SENSE OF URGENCY

Wartime analysis uses combat data to solve problems in which the time to decision is measured in days if not in hours. To be sure, sometimes the situation at hand may have been building for several months, as were some anti-submarine warfare problems described by Morse and Kimball (1951); but to a man Blackett, Jones, Dorfman, Koopman, and the others described by Thomas (Chapter 1) transmit an overwhelming sense of pace and urgency.

SOME UNCERTAINTIES ARE ELIMINATED

Wartime operations analysts emphasized the ideals of the scientific process. This was partly because of the professional roots of many of them, but also because their proximity to their clients, the combat leaders themselves, allowed the problems to be more narrowly defined. In war some uncertainties are eliminated. One knows the scene of action, if not the precise terrain or sonar conditions, the orders of battles and weapon capabilities, the tactics being used, the personalities of the commanders, and the states of training and morale. Missions and objectives are generally or precisely known.

SIMPLE MODELS PREDOMINATE

Thus the scientific approach: establishing an hypothesis in the form of a new tactic which can be tested and the results measured (as for instance, bombing results after the infusion of special training, or U-boat kills after a new depth charge setting was incorporated into ASW aircraft tactics).† The "model" is often nothing more than an hypothesis of some cause-and-effect relationship. In some instances the operations analysts reach hardly more deeply into their tool kits than for basic probability models (normal distributions, Poisson processes, and the like) or phenomenological models (such as the radar equation). In other instances the model was an extraordinarily artful description of the process which explains

†*Morse and Kimball (1951), pages 55 to 59.*

why something is happening, and therefore, what to do about it. See Thomas's description (again, Chapter 1) of World War II analysts' incisive mathematical model of the relationship between convoy size, the number of escorts, and the losses sustained, which led to a reversal of policy regarding the size of convoys. In most instances immediate problems were solved by simple, *ad hoc* models or constructs, which may have been no more complicated than arraying data in an imaginative way.

NATURE OF THE DATA: REAL BUT ROUGH

The existence of combat data should not lead to an overestimation of its precision and value. Wartime analysts emphasized the need for personal presence and observation to understand not only the nature of the data but the nature of the operation. Morse and Kimball's famous "hemibel thinking," the search for a threefold difference between postulated and observed results, was based in part on the coarseness of the data, in part on a compelling desire to find big areas of improvement.†

WARTIME MODELING IS A LOOK FOR THE BASIC PROCESS

The wartime modeler begins with observed results using whatever incoherent, unstructured data that may be available. Oversimplifying for emphasis, in wartime one works from the data backwards to find the process, whereas in peacetime battle planning one works forward, conjecturing the model processes and then generating field experiments to corroborate them.

Weapon Procurement

THE NEW ELEMENT: SYSTEM COSTS

The third milestone of model development was the introduction of systems analysis: the use of science to help choose between alternative weapon system candidates. The new element was cost, and the aim was to choose on the basis of cost-effectiveness by astutely blending cost models and effectiveness.

WEAPON COMPLEXITY DEMANDS ANALYSIS WITHOUT PROTOTYPES

Considering the fast-rising cost of developing new weapon systems and the long development and procurement lead times, it was inevitable that some disciplined form of cost and effectiveness forecasting would be undertaken. Nor could new weapons -- most notably ballistic missiles with nuclear warheads -- be field-tested in the accustomed way. It simply was no longer possible to build a few prototypes and try them out. Experimental hardware might

---

†*Morse and Kimball (1951), page 38.*

help resolve the initial uncertainties later, but by then it could be too late to reverse a decision.

The modeler cannot know with precision the future combat environment for his new weapon. For a new warship that will not enter service for five years and will be operating well into the next century, inevitably there is a debate over the characteristics of the future opposition, the friendly force in company, the geography, and even the mission. What makes weapon procurement analysis effective is the power of cost and effectiveness *comparisons*. It is much less demanding to ask for the ranking and relative merit of two or more competing systems than to ask for their absolute performance in battle.

FUTURE ENVIRONMENT: ANOTHER CAUSE OF UNCERTAINTY

MODELS ARE DESIGNED FOR COMPARING CHOICES

There is, however, a great imperative associated with the criterion of relative cost-effectiveness, namely that the weapon systems be compared on an equitable basis. In Chapter 8, Englund discusses further the problem of establishing the right criteria for fair comparison. To illustrate with two examples:

EQUITABLE COMPARISONS

- One system will probably perform better in one combat environment than another. A single-mission weapon system, such as an ASW frigate, will probably look more cost-effective when only the ASW mission is considered, whereas a multipurpose AAW missile-armed destroyer equipped with an ASW suit would look better when evaluated in a multi-threat environment.

- A weapon hardly ever fired can influence the course of battle. As the war for the Falklands showed, a high-altitude AAW missile that is known by the enemy to be deadly will keep the attacking aircraft low and subject to the two lesser evils of (1) reduced bombing effectiveness and (2) other less potent AAW point defenses. The second-best weapon sometimes shows in the record books as the major killer because the best system was too tough to confront.

CAUTION AGAINST ATTEMPTS TO TRANSFER RESULTS

Special techniques are almost always used to evaluate the candidates. Model results that are fair for purposes of comparison are most times non-transferable for battle planning or force sizing purposes. Analysts shudder when models and results that were designed for the purpose of establishing weapon preference are then used heedlessly to determine how many of the weapon to buy.

Force Sizing

The fourth milestone was the application of models to assist in whole-force assessments. Two problems resulted, the lesser of which was that the models involved were

EARLY ATTEMPTS TO ADD DETAIL AND REALISM

necessarily macro in scale. Treating this change of scale as the essential difference between either battle, wartime, or weapon procurement modeling, military operations researchers struggled to add detail to these models. We were urged on by military men who wanted to capture more of the complexity of a large-scale campaign, and abetted by a Secretary of Defense who said that analytical methods would determine the sufficiency of our force levels. We built successively more complicated models, traded verisi-

WHICH FAILED

militude for opacity, and in the end failed to achieve anything approaching our customers' expectations.

Along the way, the second and more fundamental problem dawned on the military OR community. In all the previous endeavors, a relative scale was enough. We asked only for

TOO MANY VARIABLES

a source of improvement: better battle plans, better tactics, and better weapons. By undertaking the force-level problem we asked ourselves a question that entailed an answer in absolute terms: how much is enough? It was a problem different in nature as well as different in scale. Moreover the client naturally was interested in the

AND UNPREDICTABLES

effects of small changes in force levels: even a 5% change in total force capability involves the expenditures of billions of defense dollars. But it was not possible, even in theory, to project enemy force levels and capabilities, the scenes of action, nor even our own future capabilities with adequate precision.†

GOOD PRACTICE

In the face of the great difficulty associated with force-level determination in principle, what practical measures may be taken in analyzing such problems? There are three.

FORCE MIX INSIGHTS EASIER

The first is to distinguish between force-level determination and force mix determination. Dealing with

---

†A recent research effort by the Center for Naval Analyses compared the inputs of a 1967 study of a 1976 scenario with the <u>actual</u> circumstances in 1976. Two of the worst projections were (1) our own force levels and (2) our own new weapon performance predictions.

the force mix within any given level is a much more tractable problem. Whether there is a better mix, say, of fighters and bombers; or of air, surface, and subsurface ASW platforms; or of the extent and characteristics of mobility components to incorporate into ground forces are less prodigious questions than how much force *in toto* to maintain.

ESTABLISH TRENDS WITHOUT FIRM END GOALS

The second approach is to hold in abeyance the question of adequacy in some absolute sense and wrestle merely with whether more total force capability or less is wanted: treat the question as one of direction and pace and defer the question of when to stop building or cutting until more is known. Warships authorized today will not be commissioned for five years or so. Everyone need not agree that the size of the Navy ought to be 500, 600, or 700 ships in order to agree to increase the inventory above 400. One can determine the ultimate level when the possibility is real sometime in the future.

EXTRAPOLATE FROM A "CAPABILITIES ASSESSMENT"

The third approach is to focus more attention on current force capabilities. When there is agreement on the adequacy of current forces to support current missions, then a set of benchmarks will have been established (a) to estimate the future force objectives toward which to build; and (b) to estimate the areas of greatest need (or surplus) and thus pace the rate of change in the force mix.

MODEL SIMPLICITY ACCEPTABLE

It should now be apparent why relative model simplicity makes sense for force planning. The unpredictables of that future environment are so wide-ranging that one suffers no great additional penalty in accuracy by keeping the model uncomplicated.

FORCE-LEVEL MODELS GIVE (ONLY) "IF-THEN" DATA

In effect one exercises the model in order to make if-then statements: if such and such a set of circumstances (explicit and mentally assimilable inputs) obtain in, say 1995, then such and such outcomes (explicit outcomes) result. Neither the inputs nor the outputs are credible in a literal sense. But they are the best information one can have.

EMPHASIZE RAPPORT WITH CLIENT

Since there is no hard evidence even for the inputs, there can be no corroboration. The logic of the model is its sole virtue. It is almost mandatory that the analyst work with the client or a trusted deputy to transmit what the results do and don't represent. To succeed the *model* should be designed with that client interest -- not to say suspicion -- in mind. We do not entirely mean "the simpler

...AND TRANSPARENCY, REPRODUCIBILITY, FLEXIBILITY

the better." Great errors have been made by reducing a model until it no longer captures the right processes. But transparency, reproducibility, and the flexibility to accept different inputs based on different assumptions are always desirable model attributes, and for force-level decisions are indispensable. In Chapters 2 and 9, Martin and Brown are effective in amplifying on this point.

Human Resource Planning

The growth in military models was paralleled by the development of manpower, personnel, and training models. These help determine the appropriate numbers and skills of officers, enlisted men, and civilians in the military establishment, and the amount and type of training needed. The contribution of these models is often overlooked. Their wide variety and complexity are discussed by Rostker (Chapter 6).

"ONE-SIDED" MODELS: NO ENEMY COUNTER-MOVES

One salient characteristic of human resource modeling is their one-sidedness. Whereas the preceding models had to account for a thinking, resourceful enemy, the client has more control over his manpower planning. His enemy is nature -- demographics of recruits, for example, which are known, or the amount of training required to impart a skill level, which is measurable. To be sure, there are trends that cannot be ignored, such as the increase in weapon complexity; nor known, like the future state of the economy as it affects recruiting. But even these are more predictable than what new weapon the enemy may design.

...WHICH PERMITS A LONGER LOOK AHEAD

Thus it is possible to lay manpower plans ahead over time, incorporating dynamic changes. While a warfare model may be thought flawed if it claimed to treat many action-reaction steps, a manpower model is deficient if it fails to explore the longer-term consequences. For example, during a force buildup there will be a surge of manpower accession requirements and a shortage of petty officers. The appropriate actions are transitional: in five to ten years the new accessions will move through the system and provide enough petty officers and may even choke off promotions with a "hump." Mundane as this example may be, it illustrates the obligation of the manpower analyst to design a model that displays the long-range consequences of near-term actions.

SHORT RANGE IMPLIES GREATER DETAIL

The macro manpower planner can use the law of averages, projecting mean values for broad, long-range planning. There are two cautions to add. As the horizon shortens, detail must be added. Shortages will occur in individual ratings when the population as a whole is perfectly healthy. Second, manpower and training requirements presume a knowledge of force projections, both in the aggregate and in technological detail. Those projections are always in flux, and the client must know what specific force projections were used.

FORCE PROJECTION EFFECTS

In contrast with the other models heretofore, human resource models are also used extensively for *management*. Recruiting quota, manning allowance, and personnel assignment recommendations flow directly from these models and are often acted on with little or no alteration. In Chapter 6, Rostker writes that a management orientation is a salient property of much human resource modeling.

MANAGEMENT ORIENTATION

Logistic Planning

Logistics encompasses the maintenance, supply, and transportation of the resources of warfare. Logistic modeling is sweeping in its application to:

- Maintenance and repair facility capacity and location.
- Design for maintainability and repairability.
- Purchase, stockage, inventory, and data handling for spare parts, ammunition, fuel, food, medical supplies, and construction materials.
- Transportation resources and allocations, including strategic mobility.
- Communications traffic, including the placement of message and data transmission systems.

Logistics having been recognized as an enormous part of successful military planning in World War II, special modeling tools were developed in the decades that followed. Among the operations research models that have been applied to material support problems are: linear and dynamic programming; network models; queueing models; inventory models; reliability models; and decision theory and decision trees. Many times prescriptive models are used to calculate optimal allocations of purchases, inventories, or distributions. The impact of operations research on logistics efficiency has been far-reaching and salutary. As with human resource models, logistics models are

MUCH USE OF STANDARD OR MODELS

USED FOR MANAGEMENT

are exploited not merely for investigation, but for the management of operations.

BUSINESS METHODS MAY RESULT IN WRONG OBJECTIVES

In recent years the danger of basing plans upon some set of *peacetime* criteria has become evident. As Drezner and Hillestad make clear in Chapter 7, businesslike standards of peacetime efficiency may be altogether unacceptable when vulnerability to enemy destruction or interdiction enters the picture: centralized supply or maintenance activities may be the wrong approach under wartime conditions. Support planners are rethinking their decision criteria and struggling to develop models that do justice to the enemy threat. The result of planning for a wartime environment should be better answers and insights that are more closely tied to the combat arms they support.

TIGHTLY KNIT MODELS WILL NOT HANDLE NEW CIRCUMSTANCES

Not only will material support modelers be forced to struggle with *present* battle planning, but they have to contend with the vagaries of future force projections. Unforeseen changes in the center of gravity of deployed forces, as in the Middle East at present, invoke changes in basing and strategic mobility plans. Models can derive fairly accurate estimates of current requirements, but they cannot specify the additional slack capacity to add for unanticipated contingencies.

RESULTS MUST BE HEEDED TO BE EFFECTIVE

Two other problems deserve mention here. The first is that material support requirements are often given second-class treatment, even when the requirement is agreed upon. Repair parts dollars, the need for which is clearly evident by reliability modeling, are sometimes deferred to maintain weapon production schedules. Ordnance requirements are "temporarily" underfunded even when the requirement is well-substantiated. The second problem is that combat support models seldom reflect the friction of war. The tendency has been to set aside and then forget wastage, destruction, confusion, and other inevitable wartime inefficiencies. These two problems -- deliberate understocking and what might be called negligent optimism -- will create shortages if our planning is put to the test.

National Policy Analysis

Models are also used to investigate the consequences of policy decisions that have military implications. A prominent example is the extensive modeling in and outside of the Arms Control and Disarmament Agency in support of SALT (now START) negotiations, as Lieberman discusses in Chapter 10.

Other examples are the modeling and analysis (and lack of it) associated with arms sales to foreign governments, and economic models the CIA, DIA, and other organizations might use to assess the world oil supply, a grain embargo, a freeze of financial assets, and similar non-military actions with potential military consequences. Brink-of-war crisis management and mobilization gaming have been highly informative in revealing planning weaknesses as well as acclimating executives in a prehostilities pseudo-environment.

THE ROLE OF PARA-MILITARY POLICY ANALYSIS

For a long time the linkage between prehostilities actions and wartime operations was only loosely examined. For example, many command-post gaming exercises terminated with the exchange of nuclear weapons, a weakness pointed out by Martin in Chapter 2. Arguing that one should not enter a war unless he knows how to get out of it, some policy analysts contend that war termination is an important and neglected planning issue amenable to modeling and gaming.

LINKAGE WITH MILITARY MODELS DESIRABLE

The most common models for policy analysis are one-sided and two-sided operational games. In fact, in a certain sense the scenario now *is* the model. All of the precautions we have invoked about scenario uncertainties, the proliferation of decision alternatives over time, and the irreproducibility of the results of games with human decision makers are magnified. The limitations described by Brown (Chapter 9) are all even greater in the modeling and gaming of policy decisions.

RESULTS DEEPLY PROVISIONAL

E. GENERAL PROPERTIES AND PRACTICES

"Any set of rules for obtaining models could have only the most limited usefulness at best and at worst might impede the development of the required intuition.... Modeling is an intuitive process for the experienced."

-- W.T. Morris, *On the Art of Modeling* (1967)

The Invisible Influence

We have written a great deal about how to look for model limitations. We do not want to understate their strengths. There are models that are well-shaken down and capture the essentials of phenomena superbly. Why, then, are the results of good models exercised by experts sometimes ignored? Some conclude that this is "political" influence intruding. Yes, of course, but that is another

MODELS INFLUENCE DECISIONS INDIRECTLY

ELIMINATING POOR CANDIDATES

FOCUSING THE DEBATE

CUMULATIVE EFFECT

way of saying that models cannot substitute for reality and analysts cannot substitute for decision makers. The modeling process may have done all it should by eliminating a host of choices that were non-competitive. Those that remain can be too close to call, considering the uncertainties: say within 5-10% of each other by the decision criterion. Weapon selection and force sizing decisions both lie in this realm. Model results may be quite persuasive as to the direction the defense budget ought to take, but they will never be conclusive in determining the right defense budget size to within $5 billion. In Chapter 9, Brown addresses this in greater detail, and much the same philosophy is expressed by Englund in Chapter 8 regarding weapon procurement decisions. In many cases the cumulative impact of modeling efforts over time is the single most important influence in modifying preconceptions, filling in crannies of ignorance, focusing the debate on the real issues, and structuring a more rational decision process.

Multiple-Model Research: Horizontal Interaction

1. COMPETITIVE MODELS CROSS-CHECK

In lengthy studies or protracted analytical research, analysts often run similar problems on different models and compare the results. This device serves to check model properties, processes, and algorithms when no real-life test is possible. When there are wide swings in comparative results, there is naturally cause for concern; but the fix is not, as has been proposed periodically, to choose one model, refine it, and discard the rest.

2. COORDINATED MODELS OFFER LATERAL ENHANCEMENT

The above procedure serves as a check on results by, as it were, a pseudo-corroboration. A different objective may be to do multiple-model research as a deliberate effort to refine model results. This second approach uses models in well-designed, coherent combination under a single leadership. (In the first instance the client may have asked two or more agencies or contractors to run their models in isolation and competition.) Here, for example, a standing analytical AAW model may be developed and maintained at the medial level. It uses mean value inputs and is designed to treat a wide variety of parametric inputs. Results of micro-model computer simulations of intercepter-bomber actions, SAM-ASM missile actions, and electronic warfare effects provide inputs to the medial model. War games provide checks on the medial model for command-and-control realism. A prescriptive model's "optimum" results can suggest better ways to distribute the forces and serve as an indication of their maximum potential. Test-range results provide reference-point data to benchmark the simulation inputs. The leader of the coordinated model effort maintains for his clients (there would probably be several interested commands) a master plan of tests and research experiments that are to be conducted using the several mutually supporting models.

Another example would have as its focus a fleet exercise, the purpose of which is to test a new tactical formation. Since the exercise itself could measure only a few outcomes, a great deal of analytical modeling and war gaming should have preceded the fleet test to enhance its value.

To be effective these coordinated efforts must enjoy continuity of leadership and stable modeling resources.

Model Hierarchies: Vertical Coordination

A strong approach to battle modeling -- and often to human resource and logistic modeling -- is via a multi-tiered or hierarchical family of models, with the model properties tailored to suit the level in the hierarchy. In this hierarchy, the micro, medial, and macro model operators exercise their models independently but cooperatively. For the Army's system and hierarchical structure, see Payne (Chapter 4).

HIERARCHICAL APPROACH SUPERIOR

Model results of each tier service each other both upwards and downwards. As to upwards servicing, there may be a significant amount of data processing to fit the model at the next tier. We stress the need for data compatibility: the outputs of the next lower tier must fit. Assumptions and circumstances must be consistent. Interface problems must be reduced in advance by cooperation between modeling teams. If there is no cooperation, there is a major risk of incompatibility: data and results that simply cannot be transferred.

IMPORTANCE OF COORDINATION

Nevertheless total compatibility -- data and results that fit automatically as inputs -- is so elusive that efforts to integrate and automate a single computer model from micro through macro level will probably fail, and equally important, will lose transparency and flexibility. It is better to keep the inputs, models, and results distinctive so that the client can inspect, understand, operate, and influence them at each level.

BENEFITS: CLARITY, EASE OF MODIFICATION

The other benefit of an hierarchical approach, the benefit of downwards servicing, is easy to overlook. The

ADVANTAGE OF DOWNWARDS SERVICING

larger-scale model's results help to focus the smaller-scale problem. For example, the first simple theater-level analysis of shipping protection in the early 1960s brought out some favorable characteristics of one of our weapon systems that led to a decade-long series of medial model refinements and a magnificent program of at-sea performance tests of the weapon system itself. The macro results help to identify the opposing forces and likely battle sites. The medial model results pinpoint the sensitive tactics and force deployments which merit more detailed examination. It is easy to forget how interest in many vital issues first arose through a macro perspective.

On the Misappropriation of Model Results

MODEL RESULTS NOT TRANSFERABLE

One of the frequent and insidious mistakes made by even experienced policy makers is to adapt the results of models for a purpose other than that which the analyst intended. The threat of this is the bane of all modelers. As an example, we have already indicated how weapon selection modeling uses techniques for evaluating relative merit, the results of which are not transferable for force sizing. For the skeptic we have included Chapter 12 by Battilega and Grange to be quite specific and technical on this subject.

On Data Banks

GENERAL-PURPOSE DATA BANKS HAVE NOT WORKED WELL

Since every model requires input data, the idea is frequently heard that single consolidated repositories of data to serve many models will not only simplify the search for data but eliminate the temptation to pick and choose data for inputs selectively.

General-purpose data banks have been tried. They have been plagued with difficulty. The data bank librarian must know how his data will be used, and the modeler must know rather precisely what the data represents. To obtain a value as elementary as the appropriate passive sonar detection range, one not only needs to establish the obvious conditions (such as the sonar system, target submarine hull types, and oceanographic and bottom properties) but also (for most questions that matter) shipping density; whether the sonar operator was alerted or not; the depth of the sonar and target; the speed of the sonar and target through the water; whether the detection was under controlled or free-play conditions; and conceivably even the state of training, detection bearing, and target

AN EXAMPLE

aspect. At one time or another, each of the above have been major influences in determining the mean and standard deviation of passive sonar detection range. This is not to say that the requisite input data cannot be found. The point we wish to make is that mating data with a model is a delicate process.†

On Modeling Organizations and Personalities

Given enough time, debate, and investigation there would be a high degree of consensus among analysts as to the best model to apply to a problem. In point of fact, the selection of a model often depends on the personality, education, and experience of the analyst who builds and uses it. A mathematician may look for a model long on Poisson processes or Markov chains, choosing from among his familiar tools. A statistician might fit the problem into a regression model. An economist would tend to refine his cost models and give effectiveness short shrift. Most junior military officers are instinctively drawn toward a computer simulation or a war game for "realism." A senior officer or defense official with an analysis background will work out some simple mathematical relationships in the belief that he already knows which variables matter. We all do best with what we understand best. It would be folly to ask the economist to use the best model if the best turns out to be a war game. If you need to understand the reason a type of model was used, find out who selected it. This, of course, has absolutely nothing to do with the integrity or objectivity of the person.

INFLUENCE OF PERSONALITY

The same can be said of OR organizations. Each contractor has a reputation for special skills in certain kinds of modeling and analysis. Each company's inadequacies are usually also generally known. In-house organizations have their capabilities and limitations, too. For best results you should not ask the Naval War College to build an analytical model, but no organization designs and conducts better naval war games.

UNIQUE CONTRACTOR EXPERTISE

Periodically, an alarm goes up because certain contractors seem to have an inside track on the study of

---

†Also see Rostker in "Human Resource Models," Chapter 6. The level of aggregation is an early decision in modeling. Either the data must be available or the model must be adapted to take what data there is.

similar problems again and again. Widespread competitive bidding is thought to afford opportunities for new talent. But model builders are not carloads of wheat. Contractors of analysis are offering brainpower of a particular kind. The value of having confidence in a certain modeling skill is hard to price. The quality of the contractor proposals is a poor gauge of the quality of the product, whereas the past track record of individuals working in their area of expertise is a very reliable indicator. Clients want the freedom to keep using organizations that do well, and shed ruthlessly those that fail.

For better or worse, the selection of the right modeler will in practice have much to do with the selection of the best model.

On Models and New Ideas

EXPERTS ON SUBSTANCE HYPOTHESIZE . . .

ANALYST'S MODELS TEST THE HYPOTHESES

Somehow the notion circulates that models will help to discover new tactics or the need for new weapons. It is true that models reveal shortcomings, some of which may have been unsuspected. But generally the tactician or the technologist is the creator: he hypothesizes the new tactic or weapon. Then the analyst creates a model to test the hypothesis.

To borrow from Milton Weiner†, often it's not "Let's model it so we can understand it," but "Let's understand it so we can model it." If the Army wants to explore the advantages of maneuver warfare using helicopters, then a model built upon the processes of firepower and attrition (Lanchester equations in the extreme) will be a poor vehicle to do so. A danger inherent in detailed, complicated computer simulations of combat is that they reflect current doctrine; their ability to intimate radically new tactics is limited. But the analyst may see things while he's developing the model. That is why an officer analyst who has field experience is so valuable: his talents as tactician and modeler will reinforce each other.

SPECIAL VALUE OF THE OFFICER ANALYST

On Model Stricture

As we have said, during the 1960s the military systems analysts suffered a bitter foretaste of the ways complex simulations could constrain imagination and creativity when they tried to solve problems by using the brute-force

---

†*Quoted by Martin in Chapter 2, page 108.*

capacity of the computer to add more and more inputs and make more and more runs. Beyond a point, adding detail for the sake of realism becomes self-defeating.

A MODEL MAY BECOME TOO LARGE FOR THE OPERATORS TO GRASP

With the still-expanding capacity of the modern digital computer, a different hazard of very complex computer models is upon us according to some, a hazard that might be called the stricture that the model imposes upon the user. This tyranny over the human thought process takes two forms. The first and lesser way that a computer model may constrict thought stems simply from the size and complexity of the system. Consider a large-scale information storage and processing system. A small group of experts can easily create a program used by thousands of workers, few if any of whom understand the decision rules or interactions of all the components. It is easy to think of examples of computer models that were intended to help synthesize multi-component analyses, but in the end very nearly surpassed the comprehension of the operators. In time, decisions come to be made according to the computer program's instructions that no one any longer understands and which in due course should have been overtaken by change, the need for which no one is any longer able to apprehend. Even when the need is appreciated, the modification required may threaten to render the whole system inoperable. A major large-scale planning and programming system, the Navy Resource Model ("the NARM"), was such a computer program. The NARM grew so complex, so lacked documentation of many changes, and so suffered from turnover of its operator-analysts that in due course the operators were unable to justify its interactions. It had to be discarded.

AN EXAMPLE: THE NAVY NARM

MODEL'S LOGIC MAY BE MISTAKEN FOR THE LIMIT OF THOUGHT ITSELF

The second and more insidious form of constraint on thought and judgment is a possibility that exists even with the simplest models but which is more likely to be serious with large-scale computer programs. The danger is that the model logic which is, at first, seen to be superior in providing good decisions most of the time may in due course become accepted as circumscribing the real limits within which decisions should be permitted, and finally (in the extreme) be taken as encompassing the entire range of possible decisions. The better the model provides, for example, systematic and consistent decisions in a desirable way that individual storekeepers or pay clerks cannot match, the more the model's algorithms may come to be taken as the limits of acceptable and proper decision. Military men know the importance of doctrine for cooperative action, and that training to doctrine will

ANALOGY WITH DOCTRINE

produce better than random results most of the time. But we also know that doctrine enforced in the extreme can be limiting and stultifying.

It is the same with a large computer system which substitutes decision rules and algorithms for individual judgment. An operator may be permitted no more latitude for initiative than to file, retrieve, and manipulate according to the computer program. Here it is not a question of his understanding what is permitted; he understands full well. The danger is that his capacity will atrophy for *conceiving* solutions or operations beyond those which the computer program allows. In Chapter 2, Martin describes how the early DoD computer models of nuclear war later inhibited thoroughgoing nuclear warfare planning because the models were designed to portray only simple, one-time weapon exchanges.

POSSIBLE THREAT TO CONCEPTUAL POWERS

Hayes and Abernathy (1980) have asserted that the phenomenon has even crept into managerial decision making: "During the past two decades American managers have increasingly relied on principles which prize analytical detachment and methodological elegance over insight, based on experience, into the subtleties and complexities of strategic decision."

Even so, it would be easy to exaggerate the hazard of the model running the man. The well-known ingenuity of the American soldier to "pinball" a field exercise by evading the intent of the game rules seems to go on unabated, and the younger generation knows how to sneak in an advantage over their favorite video games. But some distinguished minds have concerned themselves with the mentally delimitating effects of large-scale computers: that which cannot be programmed, they say, becomes treated as irrelevant. Moreover, if the computer model is exceedingly complicated, the inherent limitations of programmed logic are forgotten entirely. Model stricture is an ever-present possibility.

Modeling Is An Art

TALENT, ARTISTRY, AND JUDGMENT

Implied in many of our remarks is the importance of judgment in the art of good modeling. Many of our peers' commentaries on an early draft either highlighted the artistry of modeling or shrugged it off as obvious. We want to be explicit on the subject if only because models are often taken to task for being "judgmental," as if objectivity and the scientific process substituted for creative skill. To paraphrase one commentary: "Be blunt about it: say that talent is more important than education, experience, or even dogged determination."

All model construction is heavily dependent on judgment in selecting:

- the right functional relationships
- the best modeling techniques
- the scenarios (or avoidance of them)
- the sources of inputs
- the precision or spread of the results
- what conclusions are warranted

EVEN SCIENTIFIC MODELS ARE JUDGMENTAL AND NOT FINAL

No model is above challenge. Confidence in a model is heavily a matter of judgment, even when the model is subject to thorough corroboration by experimentation. Years ago the mathematician Henri Poincaré said, "With four free parameters I will fit an elephant. With five I will make his trunk wave."

Good models and good analyses result from:

- knowledge of the tools and the craft of the OR trade;
- experience on the job;
- experimentation and tests with field data and other models;
- knowledge of history, physical science, philosophy, and social science;
- knowledge of military operations, logistics, personnel, or state craft;
- the ability to parse a problem down to its essence: to "get to the bottom of things";
- and the ability to synthesize the diverse parts of a problem with, as the French say, a *coup d'oeil*, and grasp the implications all taken altogether.

What Karl Weierstrass said of Nineteenth Century mathematicians is just as true of Twentieth Century modelers: "A mathematician who is not at the same time a bit of a poet will never be a full mathematician."

NOTE 1 (page 4)

These four categories are thought to be vivid and useful, while incorporating the full sweep of military models that assist in decision making.

Analytical representations are mathematical formulas: symbolic models, but excluding word models. Analytic models often use computers to process the formulas but, in our terminology, nevertheless remain analytic models.

A *computer simulation* imitates (as it is able) physical phenomena: movement, firepower, etc. Human decision making is represented by preprogrammed action rules. Decision rules may be defined probabilistically. Computer simulations which may be stopped to accept human inputs in mid-run fall between (1) computer simulations and (2) war games.

We choose to segregate *computer* simulations from *war games*, both of which are, of course, proper simulations. We acknowledge a departure from JCS Publication 1, which defines a war game as "a simulation by whatever means, of a military operation involving two or more opposing forces, using rules, data, and procedures designed to depict an actual, or assumed real life situation." But we point out that the JCS Studies Analysis and Gaming Agency publishes a *Catalog of Wargaming and Military Simulation Models* (1983) which conveys a distinction between the two common-usage categories. In any event we believe that *war gaming* implies one or more human participants.

By defining models as abstractions...for an investigative purpose or for resource management, we have omitted two major model categories:

- training models
- phenomenological models

We have excluded models *for* training -- the wide variety of simulators, board games, and field exercises used to conduct training -- partly because it is a large field of its own, and partly because training is outside our expertise. Training models may be similar to battle models, and in fact it is US Army policy that battle models for research will be designed to double as training models.

By omitting *phenomenological* models, we have foresaken the opportunity to claim kinship with these, the most scientific of models. They include models of atmospheric and oceanographic propagation, signal processing, nuclear weapon effects, propulsion and resistance, human behavior, and many more. Phenomenological models are the most elemental of our building blocks. Although we have deep associations with these models -- and indeed the early practitioners of operations research and systems analysis were the physicists, economists, biologists, and psychologists who were grounded in them -- it is better to regard ourselves as the users rather than the architects of phenomenological models.

NOTE 2 (page 6)

The literature of operations research and science is rich with struggles to deal with uncertainty. Setting aside matters of stochastic uncertainty (though they are vital and pervasive), it may be useful to include a brief discussion of uncertainties related to the application of deterministic functions. Whether the underlying determinism (explicit cause-and-effect property) is by knowledge, conviction, or presumption need not affect this discussion.

Important mathematical functions may be represented by an infinite converging series, for example:

$$\text{Sine } X = X - \frac{X^3}{3!} + \frac{X^5}{5!} - \frac{X^7}{7!} + \cdots \qquad \text{for all real } X$$

The first few terms are an approximation of the value of the function. The additional terms refine the approximation to any degree of desired accuracy.

In physics, the function $\text{Sin}_{\Delta}\ \omega t$ *may be a very accurate model* of phenomena with wavelike properties: acoustic, electromagnetic, optical, etc. In operations research the model may be a mathematical function, but typically it is *not offered as a very accurate representation* of the object being modeled. Rather, the function is asserted to have properties that capture its essence. Forms of the exponential function (e^x, $e^{-\lambda t}$, or $1 - e^{-\lambda t}$) appear again and again in the OR literature, for example. The exponential function may be represented by the infinite series:

$$e^X = 1 + X + \frac{X^2}{2!} + \frac{X^3}{3!} + \frac{X^4}{4!} + \ldots$$

When the mathematical model is a complicated function of many variables, another departure from accuracy may be acceptable. A linear approximation is sometimes substituted. For instance the input-output relationship of one variable Y may be *abstracted* in the model as e^{-Y}, and the abstraction further *approximated* as:

$$f(Y) = e^{-Y} \doteq 1 - Y$$

Of course, the greater the departure of a new value Y_1 from an initial value Y_0, the greater the error introduced by the linear approximation. Figure 3 illustrates. Let Y change from an initial value of $Y_0 = 0$ to $Y_1 = 0.5$. Then $f(Y_1) = e^{-0.5} = .606$. The linear approximation $1 - Y_1 = 0.5$. At $Y_2 = 1$, $e^{-1} = .368$ but $1 - Y_2 = 0$. Values of e^{-Y} will never be negative, but the approximation $1 - Y$ is negative for all values of Y greater than 1.

Figure 4 illustrates that when the third term of the series expansion is added:

$$f(Y) = e^{-Y} \doteq 1 - Y + \frac{Y^2}{2}$$

then the error of the series approximation is reduced, and is now an error on the high side.

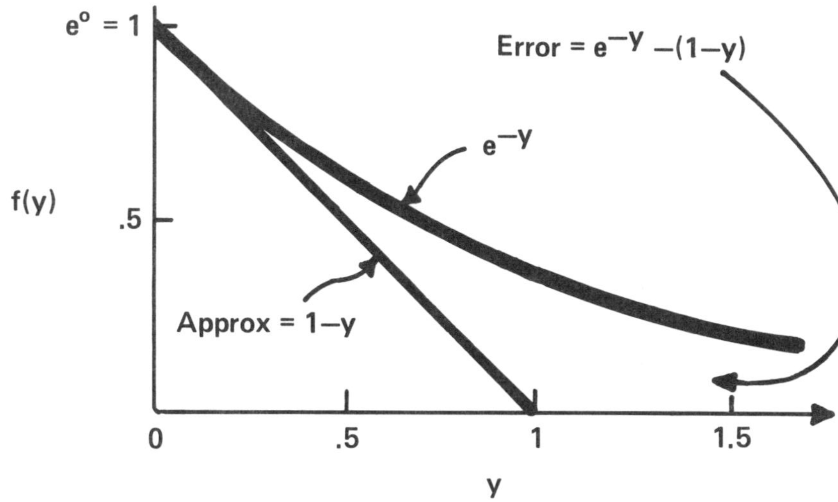

Figure 3. Mathematical Approximation by Two Terms of a Series Expansion

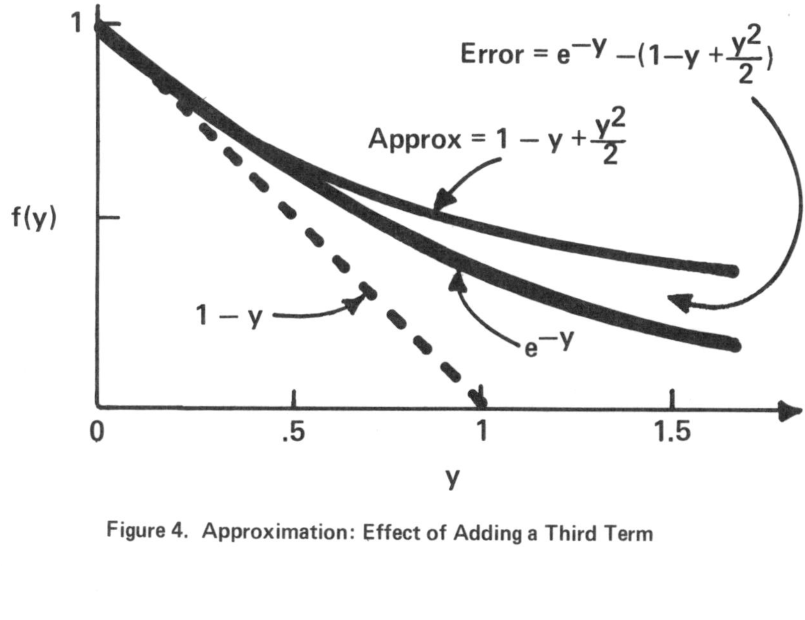

Figure 4. Approximation: Effect of Adding a Third Term

Operations research models in general deal with both kinds of simplifications: (1) abstractions from reality to allow expression of the phenomenon mathematically, and (2) approximations of the mathematical functions to make manipulations of the equations more tractable. Payne makes this clear in Chapter 14 when he discusses model validation.

It may readily be seen why an operations research model containing approximations (and many, if not most, do) will yield highly satisfactory predictions when the inputs are changed from a known set of initial conditions to a new set of conditions, but only if the inputs are not changed too radically; a model's predictive power can become less and less satisfactory as the amount of change increases. In operations research, seldom is it possible to test a model widely enough to conclude that it is good for general purposes. On the other hand, for many useful models, general applicability is not essential as long as the model's limits are not abused. In a very real sense it is not the model alone that is utilitarian, but the model in conjunction with some acceptable range of inputs.

Until now this discussion has taken the point of view that the analyst has a set of data X, Y, and Z; an associated outcome $W = f(X,Y,Z)$; and in addition knows, more or less, the functional relationship f between output W and the inputs X, Y, and Z. In practice, the shape of the function f is usually comprehended only dimly. In Chapter 1, Thomas refers to the World War II operational analyst P.M.S. Blackett's description of a procedure to follow when one has only a numerical outcome Y of a wartime operation. The analyst conjectures $Y = F(X_1X_2...X_2)$, in which $X_1, X_2,...X_n$ are all influential inputs and F is the unknown relationship. Blackett rejects the idea of deducing F from first principles. The useful wartime procedure, he asserts, is to attempt marginal estimates of Y' from new inputs X_i'.

The situation Blackett describes is depicted in Figure 5 (in one variable). There is some unknowable function F(X) of great interest. Inputs X_0 and X_1 yield corresponding outputs Y_0 and Y_1. Then within some range of values around X_0 and X_1, an assumption that $\frac{dF}{dX} \doteq \frac{\Delta Y}{\Delta X}$ will be useful.

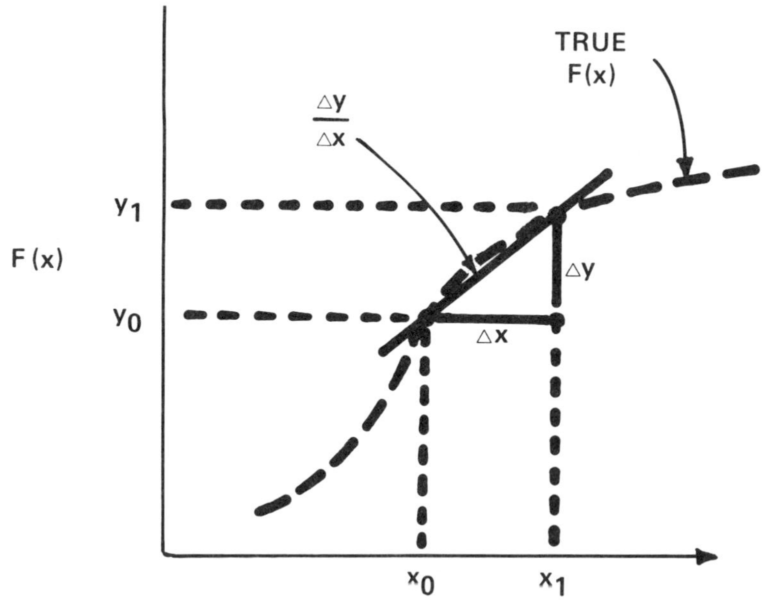

Figure 5. Estimating a Non-Linear Function with Changed Data Δy/Δx

Reference to a Taylor series expansion offers insight at this point:

$$f(X) = f(a) + (X-a)\ f'(a) + \frac{(X-a)^2}{2!}\ f''(a) + \frac{(X-a)^3}{3!}\ f'''\ (a) + \ldots$$

where $f^{(n)}$ is the nth derivative of f and a is a point in f(X). In the second term, the Taylor series f'(a) is the first derivative of the function. It is the slope -- the instantaneous rate of growth or decay at the point a. In the third term, f"(a) gives the rate of change of the slope at a, and if known, is highly useful to the analyst in indicating the shape of curvature of the true f(X) -- its rate of change from a linear relationship.

In battle model applications, the difficulty of incorporating the measures and countermeasures possible on each side increases rapidly over time. The initial conditions, a in this instance, can be specified and understood. The initial measured result, f'(a), indicates the probable battle dynamics at the outset. If the second derivative, $f''(a)$, can also be ascertained, that information will be very valuable in suggesting how the initial results (attrition rates, movement rates, etc.) will accelerate or decay. Beyond that it is risky to place much credence in battle model results over time.

Operations research modelers compromise on the use of mathematical *approximations* more willingly to the extent that they believe they have already compromised on the model's fidelity to attain a mathematical formulation. At times they may believe that though the model itself can be made quite accurate, the inputs will be imprecise. In this case they may be more willing to compromise on the model as an *abstraction* as well.

NOTE 3 (page 11)

Corroborate is a more accurate description of the process which is generally called *verify* or *validate*. Corroborate means to strengthen or support with other evidence or authority, and so is less exacting than verify or validate.

There have been attempts to distinguish validation of a model from verification. At least as long ago as 1967, Dr. Seth Bonder proposed the following distinction:

> *Validation* of a mathematical model requires that it be internally consistent...the model conclusions...arrived at by correctly applying formal rules of logic.... Nothing is implied regarding the empirical truth of the conclusions.
>
> *Verification* of a model is the process of statistically testing deductively derived hypotheses with experimental evidence...a model can never be completely verified. The decision that the model and data do not disagree is a subjective one on the part of the scientist.

On the other hand, USAF directives on validation and verification make just the opposite distinction. In practice, both words will be seen referring to experimental confirmation that a model is "right" in some sense.

One can argue that while the continuing process of model corroboration is easy to understand, it is hard to define a "corroborated" model. To verify means to prove, which is what one would like to do. To validate means to declare well-grounded or sound, which is an unambiguous assertion of legitimacy. But to corroborate merely means to make a better case than before. Precisely so. The picture of continuing provisionality and residual margin for error and improvement in most military models is an accurate one, suitably conveyed by "corroborate."

NOTE 4 (page 18)

When it comes to Occam's Razor, each authority seems to have a slightly different version. Not only the translation but the Latin original vary, as do the spelling of William of Occam's name (Ockham, Okham) and even the dates of his life (1280-1350 ± a few years). The source chosen was the *Encyclopaedia Britannica*, 1967, under Ockham, William of, Volume 16, page 858, if it matters.

Occam's principle is not simply the KISS Principle (Keep it simple, stupid) in medieval garb. It is the assertion that the simplest explanation is the correct one until it is shown to be irreconcilable with the evidence. But the roots of the principle certainly antedate William of Occam:

> [Claudius] Ptolemy [whose active scholarship was from AD 127 to AD 151] was far ahead of his time in his views of how scientific research should be conducted. In an elaboration of Hipparchus, Ptolemy said that for the best explanation of any phenomenon one should adopt the simplest hypothesis that it is possible to establish, provided that it does not contradict in any important respect the result of the observation.

(Quoted from John N. Wilford, *The Mapmakers*, Random House, New York, 1982, page 26.)

REFERENCES

1. Russell L. Ackoff, Scientific Method: Optimizing Applied Research Decisions, Wiley, New York, 1962.

2. John A. Battilega and Judith K. Grange, eds., The Military Applications of Modeling, 3 volumes, Science Applications, Inc., Denver, 1978.

3. Edward A. Bender, An Introduction to Mathematical Modeling, Wiley, New York, 1978.

4. P.M.S. Blackett, Studies of War, Hill and Wang, New York, 1962.

5. V.M. Bondarenko and A.F. Volkov, Avtomatizatsiya Upravleniya Voyskami (Automation of Troop Control), Voyenizdat, Moscow, 1977. Translation by Foreign Broadcast Information Service (JPRS 8199).

6. Seth Bonder, Topics in Military Operations Research, printed notes for course given July 25-29, 1966, at the University of Michigan. Included are chapters on "Field Experimentation" by J.W. Hazell.

7. Garry D. Brewer and Martin Shubik, The War Game: A Critique of Military Problem Solving, Harvard, 1979.

8. L.G. Callahan, Jr., ed., Proceedings of the Workshop on Modeling and Simulation of Land Combat, Georgia Institute of Technology, 1982.

9. Yu. V. Chuyev and Yu. B. Mikhaylov, Prognozirovaiye v Voyennom Dele (Forecasting Military Affairs), Voyenizdat, Moscow, 1975. Translation by U.S. Government Printing Office, 1980.

10. V.V. Druzhinin and D.S. Kontorov, Ideya, Algoritm, Resheniya (Concept, Algorithm, Decision), Voyenizdat, Moscow, 1972. Translation by U.S. Government Printing Office, 1975.

11. Admiral of the Fleet Sergei Gorshkov, "Problems with Respect to Control of Naval Forces," Morskoy Sbornik, Issues 5 and 6, 1980.

12. Robert Hayes and William Abernathy, "Managing Our Way to Economic Decline," Harvard Business Review, July-August 1980.

13. Olaf Helmer and Nicholas Rescher, "On the Epistemology of the Inexact Sciences," Management Review, Volume 6, Number 1, October 1959.

14. Charles J. Hitch, "An Appreciation of Systems Analysis," ORSA Journal, Volume 3, Number 4, November 1955.

15. Francis P. Hoeber, Military Applications of Modeling: Selected Case Studies, Gordon and Breach, New York, 1981.

16. Horace F. Judson, The Search for Solutions, Holt, Rinehart, and Winston, New York, 1980.

17. Major General Glenn A. Kent, "On Analysis," Air University Review, May 1967.

18. Bernard O. Koopman, The Logical Basis of Combat Simulation, Office of Naval Research Report N00014-67-A-0108-0001, November 1968.

19. Bernard O. Koopman, "Intuition in Mathematical Operations Research," ORSA Journal, Volume 25, Number 2, March-April 1977.

20. Imre Lakatos, Proofs and Refutations: The Logic of Mathematics, Cambridge University, 1976.

21. Lawrence J. Low, Chairman, Theater-Level Gaming and Analysis Workshop for Force Planning, Volume I (Proceedings) and Volume II (Summary, Discussion of Issues, and Requirements for Research), Office of Naval Research, Arlington, VA, 1977 and 1981 respectively.

22. Alan L. Meyrowitz, "Research Directions in Artificial Intelligence," Naval Research Reviews, Office of Naval Research, Volume XXXIV, Number 4, 1982.

23. William T. Morris, "On the Art of Modeling," Management Science, Volume 13, Number 12, August 1967.

24. *Philip M. Morse and George E. Kimball, Methods of Operations Research, M.I.T. Press, First Edition Revised, 1951.*

25. *Henri Poincaré, "Mathematical Creation," The World of Mathematics, James R. Newman, ed., 1956.*

26. *E.S. Quade, Analysis for Public Decisions, American Elsevier, New York, 1975.*

27. *Richard Richels, "Building Good Models Is Not Enough," Interfaces, The Institute of Management Science, Volume 11, Number 4, August 1981.*

28. *Omond Solandt, "Observation, Experiment, and Measurement in Operations Research," ORSA Journal, Volume 3, Number 1, February 1955.*

29. *Charles M. Sternhell and Alan M. Thorndike, Antisubmarine Warfare in World War II, Operations Evaluation Group Report No. 51, Navy Department, Washington, 1946 (especially Part II, "Antisubmarine Measures and Their Effectiveness").*

30. *James G. Taylor, Force-on-Force Attrition Modelling, ORSA, 1980.*

31. *Clayton J. Thomas and Walter L. Deemer, Jr., "The Role of Operational Gaming in Operations Research," ORSA Journal, Volume 5, Number 1, February 1957.*

32. *Frederick D. Thompson, "Beyond the War Game Mystique," USNI Proceedings, Volume 109, Number 10, October 1983.*

33. *Peter W. Zehna, ed., Selected Methods and Models in Military Operations Research, U.S. Government Printing Office, 1971.*

34. *The Journal of the Operations Research Society of America, Volume 19, Number 5, September 1971 (devoted to OR professional standards).*

35. *Research Analysis Corporation, NATO Combat Capabilities Study, Volume VI (Rates of Advance of Theater Forces), TAC-CR-56, June 1972 (abridged in Appendix II of GAO Models, Data, and War, 1980).*

36. *Studies Analysis and Gaming Agency, Catalog of Wargaming and Military Simulation Models, JCS, 1983.*

37. *U.S. Army Logistics Management Center, Defense Logistics Studies Information Exchange, Ft. Lee, VA 23801, Department of Defense Catalog of Logistics Models, January 1983.*

ABOUT CHAPTER 1

In these 15 chapters we will illustrate and amplify the principles of good modeling we established in the Overview. We start with Clayton Thomas's examination of wartime applications. He illustrates in more detail than the Overview why models used in wartime have characteristics all of their own.

Drawing from examples of important wartime analyses, Thomas shows that highly successful work owed little to sophisticated models, either mathematical or computer. The Operations Research method, as pioneered in World War II, depended strongly on a *scientific outlook*. Essential to the successes were skepticism, energy and ingenuity in digging for sparse "facts," and deep thought and imagination in how to measure the effectiveness of an operation. These activities, as in scientific investigation generally, often led to model building, but the models tended to be simple and *ad hoc*. On the one hand, the war itself served as the "overall model," the ultimate "test range." On the other hand, in the midst of war, analysts had to deal with low-quality data -- real but "dirty" -- which was confused by the complexity of combat, the variety of operations, and deliberate activities of the enemy to confuse and disinform. This and the obvious prominence of leadership, morale, courage, and luck all serve to remind us of the provisional nature and specific applicability of any model of conflict.

Thomas concludes by encouraging us to plan in peacetime to be better prepared to apply wartime methods and models.

Models and Wartime Combat Operations Research

by

Clayton J. Thomas

INTRODUCTION

The success of combat operations research in World War II gave rise to the first large-scale -- and continuing -- institutionalization of operations research. From that early work, one can trace an evolutionary course to the spread of systems analysis and the development of such tools as large computer models. It is interest in these large models and in problems associated with their proper use that has led to this monograph and its chapters on battle models, weapon selection models, force sizing models, logistics models, etc.

Because the development of models has played such a prominent role in military operations research, and because wartime combat operations research itself gave birth to the field of military operations research, one might very well expect to find some or all of the following:

- An explanation for the great success of World War II operations research, primarily in the use of combat models.

- Continuous development over time of progressively improved combat models, with correspondingly marked advances in the operations research of the Korean and Southeast Asia conflicts as compared to that of World War II.

- A community of busy modelers dedicated in peacetime to the production of improved combat models to be ready for use in the next war.

Actually, however, as the present chapter underscores, there is surprisingly little evidence for any of the above "research hypotheses." Instead, a review of operations research as conducted in World War II, Korea, and Southeast Asia suggests the following:

- The "scientific outlook" of World War II operations research pioneers was central to their success. Their primary emphasis was on devising measures of effectiveness and collecting data to find out how "the war was going" and how to improve operations. Often data plots or tables gave key insights without an explicit model. Statistical analysis, standard probability models, basic kinematic relationships, etc., were all important. Very few of the successes required anything beyond the above, though some more advanced models were quite useful and others stimulated insights or postwar developments.

- The operations research work of World War II, Korea, and Southeast Asia exhibited striking similarity but little progressive improvement in quality. Though the menu of available techniques increased with time, much that had been learned in World War II was forgotten and relearned in later conflicts.

- In peacetime there has been little deliberate development of improved combat models for future use. At the outbreak of hostilities one takes any useful available model, whatever its original purpose. However, most large models developed in peacetime as surrogates for war are needed less when there is "a war on."

The observations above lead directly to the title, scope, and emphasis of this chapter. Its stress is on the principles that underlie good use of models in combat analysis. While it gives examples of combat models, its aim is not cataloging or training. Its emphasis, rather, is on "scientific outlook" that sees models as tools to be used when appropriate. The value of this reminder of fundamentals may well go beyond the domain of combat modeling.

GUIDE TO THE REMAINING SECTIONS

The remainder of this chapter provides background and "reminders" that help to explain the role of models in wartime combat operations research.

The first section sketches some of the differences between wartime combat analysis and peacetime analysis. Fighting a war is not the same as fighting a budget battle. Sometimes, as in Korea and Vietnam, we try to do both at once, fighting in Asia as we budget for European forces. To the extent that we are concerned with an ongoing war, however, we have a powerful focus. The war provides our problems and the urgency of solution. The war itself may be viewed as "the model" or "the field exercise" that dwarfs our efforts. Wars are not "instrumented" for analysis, so that "wartime data are dirty -- but real." Finally, the urgency of war may lead (though not always) to an infusion of scientific talent such as that which gave rise to World War II operations research.

The next section briefly reviews the methods of wartime combat operations research. Using a scheme of Dorfman (6), it discusses the five important aspects of problem perception, problem formulation, observation, analysis, and presentation. These are not "one-time" steps to be accomplished in a simple sequence, though it is easier to discuss them that way. The focus, however, must be on the problem and its solution, which is not a report but an effective recommendation for action. Many of these steps require the operations research analyst to be not only in the "theater" but actually in the "field." Where do models enter Dorfman's scheme? Problem formulation typically involves the construction of an hypothesis, and the hypothesis may take the form of "a simplified mathematical model in which the measure of effectiveness is expressed as a function of the major influencing factors."

The third section, COMBAT DATA, discusses some of the difficulties encountered in getting "good data" in the observation phase. It illustrates the untrustworthiness of human perceptions, which leads, in particular, to "bomb" cameras and "gun" cameras. Forms for data collection and large "automated" data bases give mixed results. On-the-spot explanations of the value of data by the using analyst are often key to enlisting enthusiastic support. Finally, this section reminds us of the "hemibel thinking" described by Morse and Kimball (19), a useful rule for screening data differences and important problems.

The next section uses some simple examples to illustrate the role of models, and it includes summary comment on some models not illustrated in detail. Some of the examples illustrate the value of data collection and presentation without much more in the way of modeling. Other examples, however, illustrate what a model can sometimes add to our understanding.

The next-to-last section gives a brief "look ahead." Developments in sensors and the attainment of "computational plenty" may contribute to marked changes in future combat analysis, such as much greater automation.

The last section is a short summary. It is also a repository for the "morals of the story" and the "lessons learned" that most need retelling.

THE SETTING OF WARTIME COMBAT OPERATIONS RESEARCH

When one looks at reports of the combat operations research performed in World War II, Korea, and Southeast Asia -- three very different conflict situations -- he is struck by how much they had in common. Each had its concern with radar performance, bombing accuracy, interdiction effectiveness, air-to-air exchange ratios, improvements in tactics, responses to enemy innovations, etc. Though the weapons differed, the reports were remarkably similar in their tabulations, plots, photographs of damaged targets, etc.

It is also very striking to note how such combat reports differ from the peacetime analyses performed in support of the planning, programming, and budgeting system. What were outputs of combat analysis reports often look more like inputs to some of the large simulations conducted in peacetime. Much that is parametrically varied in peacetime studies is known in war, and much that is the subject of anxious measurement in combat operations research is unconcernedly calculated in peacetime simulations.

Such differences were seen with special clarity by some of the talented pioneers. Thus, writing of his experiences as chief of British scientific intelligence in World War II, R.V. Jones (12) observed:

> Despite any unpopularity, I survived because war is different from peace. In the latter fallacies can be covered up more or less indefinitely and criticism suppressed, but with the swift action of war the truth comes fairly quickly to light -- as Churchill said, 'In war you don't have to be polite, you just have to be right!'

The "swift action of war" gives urgent emphasis to the "here and now," which has many implications for combat operations research. For one thing, there is more emphasis on what can be done with the weapons in hand. One looks at the allocation of those weapons and the tactics of employment with the greatest of care. Maintenance and reliability are special concerns. Another consequence, as Koopman (in Koopman et al. (13)) recalled of World War II:

> Now, it's true that during the war, costs meant very little. The simple reason is that we couldn't buy something else. It simply took too long to develop a whole new system, so we had to use what we had.

Koopman went on to point out that as costs became more important after the war, the term "cost-effectiveness" came in, economists became more important, and mathematical physicists less so.

Another consequence of being in a war is to simplify and clarify much that had been unknown or ambiguous. Hitch (11) points out how much more difficult it is to do a systems analysis of future systems, where no one scenario dominates, than it was to do operations research studies of weapons in World War II, where one knew who was fighting and much about their equipment. Consider in this light one consequence of Pearl Harbor. A scenario that might have seemed very unlikely a day or two earlier, suddenly, on December 7, 1941, became very compelling in the intensity of the interest it evoked.

This kind of learning applies not only to high-level scenarios but also to tactical details. Speaking at a recent conference, Kahn (in MITRE (18)) remarked:

> At one time I had been in the air defense business about 20 years. Many people here have been in it just as long. Not one of us thought a pilot could see a missile with his eyeballs. But one of the main defenses of our planes over Hanoi was a pilot seeing a Mach 2 missile coming at him and evading it. That was a totally unexpected tactic.

The possibility of such revelations confers on war a reality that is unattainable by studies of possible wars. Much of the spirit of Blackett (2) is conveyed by such words as "actual," "real," and "fact." Even the ability to know someday a yet unknown fact can contribute to this spirit. For example, in a wartime discussion of the effectiveness of bombing German cities, Blackett looked forward to the time after the war when certain facts would become known. In sharp contrast is our inability ever to know real facts about most of the fictional futures studied in broad systems analyses.

With this emphasis on the actual and factual, it is not surprising that the operations research pioneers hoped that their profession would be "commonsensical" and its practitioners "hardheaded." Blackett (2) described his variational approach as a "more commonsense procedure." Solandt (24), while recalling some of his wartime experiences, wrote:

> I for one hope that we shall never come to the day when a knowledge of the latest mathematical gadgets for war games, of communications theory, and so on becomes the badge of office of the operations research worker. I think the operations research worker should be known as a hardheaded experimental scientist, with a very broad vision, who trusts only what he can measure.... If everyone sticks to these high ideals, I think ultimately operations research may come to justify what seems to me the most flattering, but, unfortunately, unreal definition of it; that is, that operations research is the systematic application of common sense.

Dorfman (6) wrote of the need for an analyst to be "a hardheaded staff member," though other staff members may see the analyst as "a 'long-hair' and a theoretician." Dorfman's point was that while everyone acts on "theory," the analyst must be aware of the theories on which he acts and test them against experience in a continual reexamination.

All of this serious emphasis on reality and hardheadedness reflects the serious nature of war itself. While the onset of war may strip away many spurious scenarios and reveal the identity of the enemy, that enemy will devote great energy to conceal his movements or to actively seek to mislead us. He is out to win, as are we. Thus, any serious war constitutes a great challenge, and a review of the setting as characterized above in this section shows a particular challenge to exercise of the qualities of the ideal hardheaded scientist. World War II was a serious war in that sense, and there was a remarkable outpouring of scientific and analytic talent that led to the well-known operations research successes. As pointed out by Leckie (17), however, in his survey of American wars, World Wars I and II were exceptions to the rule in the national consensus that was evoked. Korea was not a popular war, and it was not as easy to get good analysts to go to Korea. Vietnam was, of course, less popular still, perhaps marking an extreme in this respect.

There were other important differences also. Neither Korea nor Vietnam was a "full-time," all-consuming war. President Truman insisted that Korea be known as a "police action," and Vietnam was fought without calling up the National Guard. Moreover, these were Asiatic wars, and while each was in progress there was a concern that as a nation we would be dangerously distracted from the threat in Europe, traditionally seen as more important.

Despite these differences and much more limited scope, Korea and Vietnam did elicit combat operations research in the mold of World War II. Here two interesting observations may be made. The first is to note the greater menu of techniques available in Vietnam. Some technique developments, in fact, have been inspired by combat operations research problems but completed or popularized after conflict was over. Thus, planning problems of World War II led Dantzig to work on the simplex algorithm, and interdiction problems of Korea led Ford and Fulkerson to work on network flow algorithms. The latter, and other similar algorithms, were available for use in Vietnam, and a few applications were attempted but were constrained by data limitations.

The other observation is somewhat in the nature of a "counterexample." Very often, what has been learned in an earlier conflict has not been known by later practitioners. Whitmore (27) called attention to naval studies in World War I, where Thomas Edison introduced methods much like those that were "introduced" again in World War II. Later, Whitmore (28) pointed out that even some of the better known results of World War II were not generally known to operations research workers in Korea and themselves had to be rediscovered. Perhaps such "forgetting" is a kind of "retroactive inhibition" that occurs in peace, making it necessary to rediscover many combat operations research "truths" in the stark reality of war.

METHODS OF COMBAT OPERATIONS RESEARCH

The greatest development of the methods of combat operations research came in the wartime setting of World War II. At the time there was no "modeling community" in the modern sense, and, at the beginning of the war, very little even of an "operations research community."

As Morse recalled (in Koopman et al. (13)):

> ...we were starting from scratch. We had to bring in and train people from all sorts of different backgrounds. We didn't have a supply of... Ph.D.'s in operations research to pick up and try to retrain. We had to retrain physicists and mathematicians, biologists and so on.... We had practically no background...we had to build up our theory....

Though the operations research workers in World War II initially had little experience in the analysis of combat operations, they did have an important contribution. Newman (20) characterized it as follows:

> What scientists brought to operational problems -- apart from specialized knowledge -- was the scientific outlook. This in fact was their major contribution. They tended to think anew, to suspect preconceptions, to act only on evidence.

Newman also emphasized the "mathematics of probability" as an "indispensable tool." This tool is, of course, the basis of our statistical procedures for dealing with evidence where chance effects are important. As one might expect, probability models are especially useful in combat research.

The "scientific outlook" has several important aspects. Solandt (24) emphasizes "observation," "experiment," and "measurement" as the key to good practice in operations research. Morse and Kimball (19) describe the approach as:

> ...past operations are studied to determine the facts, theories are elaborated to explain the facts, and finally the facts and theories are used to make predictions about future operations....

The above prescription does not explicitly mention "model," and though Morse and Kimball (19) give valuable discussions on probability, bombing problems, gunnery problems, search theory, Lanchester equations, and many, many other topics, its index does not list "model."

In the early days of model building in combat operations research, models were seen more as tools to be used in the formulation of hypotheses or theories. Since scientific tradition makes the provisional nature of hypotheses very clear, a recognition of the need to test models was essentially "built-in." Also, the models were simpler and far more transparent than the very large computerized models of today.

One useful characterization of the combat operations research approach that does explicitly mention models is that of Dorfman (6). For expository purposes, he divides the approach to a problem into five stages:

PERCEPTION
FORMULATION
OBSERVATION
ANALYSIS
PRESENTATION

As brought out more fully below, models may be used in the expression of hypotheses introduced in *formulation*. If so, one function of *analysis* is to use the data collected in the *observation* stage to test the model.

These five stages of perception, formulation, observation, analysis, and presentation will be used to organize the discussion in the remainder of this section. It should be emphasized, however, that what is convenient for exposition is not intended to suggest a simple step-by-step approach as normal procedure. Typically there will be iteration of the steps and feedback from one step to another. One may start, for example, with crude *observations* which lead to the *perception* of a problem.

This *perception* may lead to a *formulation* that involves more refined *observations* as a key step. Or, to go to the other end, when the *presentation* stage leads to the acceptance and implementation of a new procedure, there should be *observation* and *analysis* to see if it is working as well as expected. Keeping the importance of iteration in mind, it is now useful to say a little more about each of the five stages.

The *perception* that a problem exists may be very simple or very demanding. This stage requires the recognition that alternatives exist and that there is an option to choose. The recognition that present procedures are not sacrosanct and might be altered sometimes demands considerable creativity. Also, every operations research practitioner has the experience, on occasion, of finding that the problem he was initially asked to solve is not what he eventually comes to perceive as the most fundamental problem. Often, of course, the general nature of a problem (e.g., low bombing accuracy in the 8th Air Force in the early days of World War II) may be painfully evident, so that the analyst essentially starts with the formulation stage.

The *formulation* stage is extremely important in that this is where measurement and hypotheses (including models) enter. Deciding what to measure is critical. This is often termed the formulation of a measure of effectiveness (MOE), and it may be necessary to use more than one. Morse and Kimball (19) have an important chapter devoted to "The Use of Measures of Effectiveness." The classic paper of Hitch (10) emphasizes the importance of the "criterion problem" in operations research. It points out that "the shrewdness with which criteria ('payoffs,' 'objective functions') are selected" has much to do with the usefulness of operations research. Hitch emphasized the importance of making an MOE used at a low organizational level consistent with an MOE used at a higher level. Whitmore (28) gives an illustrative example:

> ...there is still a place for the individual piece of deep insight, which sounds deceptively simple after it is first announced, but which transforms an entire analytical effort. One that has always stuck in my memory is Glen Camp's remark that the Navy's problem in protecting merchant shipping was not undersea warfare, but overseas transport. This did not diminish the strong emphasis on antisubmarine measures, but did effectively

> set them in a wider context. It suggested studies of methods of decreasing turnaround time in ports, placing of especially valuable cargoes in the least vulnerable positions in a convoy, analysis of unorthodox high-speed merchant ships, and the like.

One danger should be remarked. Overt use of an MOE by management may lead to unfortunate results. One example comes from the 8th Air Force bombing problem. A very useful MOE for analytical purposes was the percent of bombs dropped which fell within 1,000 feet of an assigned "point of impact." This MOE was used in judging the efficacy of changes in tactics, and eventually it was also used to rate the performance of bomb groups. When group commanders learned that they were being scored on this basis, they sometimes, as Dorfman (6) recalls, "declined to release their bombs at assigned targets when visibility was only middling good." Management use may require hedges that are better ignored in an analytical MOE.

Once an MOE is formulated one may proceed to frame an hypothesis. A first step is to list, *a priori*, the main factors on which the MOE might depend. These factors may be divided into four sets according to "control." Those factors under friendly control, for instance, make up the first set. These are the factors we can do something about. It is also important, however, to consider as a second set those factors under enemy control. Here the spirit, if not the techniques, of game theory may be glimpsed in the wings. Those factors not under either friendly or enemy control, which are "uncontrollable but observable," Dorfman puts in a third set. One might say that these factors are under the control of "nature." In a fourth set are "chance" factors, which "account" for the variation in results even when all of the other factors are held constant.

An hypothesis then makes a testable assertion about the role of the factors. It is here that models enter. As Dorfman (6) puts it:

> The hypothesis takes the form, quite frequently, of a simplified mathematical model in which the measure of effectiveness is expressed as a function of the major influencing factors.

When such an hypothesis is tested, there are various possible analytical results. The postulated function may be good enough for practical purposes. If so, one may estimate the numerical value of important operational parameters. It may even turn out that some of the factors included are not very important, so that one can find a simpler model. On the other hand, it may turn out that additional explanatory factors are needed. These analytical results, of course, come in the *analysis* stage. What is important in the *formulation* stage is the recognition that a useful hypothesis must lend itself to such analysis.

The *observation* stage is of such importance that the entirety of the next section of this chapter is devoted to combat data. It must suffice here to note its criticality. Experience has shown that human observations and recollections must be very carefully checked and often supplemented. *Observation* is typically onerous, often dangerous, and as Dorfman notes, takes the analyst away from his desk, but it is indispensable.

The *analysis* stage, strictly speaking, tests hypotheses against the observed data, and if the fit is good, estimates the value of operational parameters.

In actual fact it is often difficult to so confine analysis. It often leads to reformulation of hypotheses, modification of models, and may point the way to other data requirements.

In this stage it is often necessary to estimate the individual effects of several factors that together influence the outcome of an operation. A classic statement of one approach is given by Blackett (2). He had little confidence in our ability, *a priori*, to construct a useful overall model of an operation. He suggested, rather, a "variational approach" that estimates how operational results might change if we change some of the variables under our control. He wrote:

> The more commonsense procedure is to abandon the attempt to construct from 'first principles' a complete imaginary operation *something like* the real one under investigation, and to replace it by an attempt to find, both by experimental and by analytical methods, how a real operation would be altered if certain of the variables (e.g., the tactics employed or properties of the weapons used) were varied.

> Suppose the result of any operation of war is denoted by quantities Y_1, Y_2, etc., called the yields. Then these yields must be considered as functions of a large number of operational parameters or variables $X_1, X_2, \ldots, X_n$. Some of these can be given a quantitative measure but some can only be expressed qualitatively. Any attempt to find the form of such a function:
>
> $$Y = F(X_1, \ldots, X_n)$$
>
> from first principles is useless, in general, owing to the complexity of the problem and the non-quantitative nature of many of the variables.

Blackett then goes on to sketch his procedure for using the result Y of some past operation (characterized by values $X_1, \ldots, X_n$) to estimate the result Y' of an operation characterized by $X_1', \ldots, X_n'$. His method may be visualized geometrically as fitting a hyperplane that is tangent at the point $(X_1, \ldots, X_n)$ to the multidimensional surface that represents the yield Y. Blackett cautions against using this method if the new X_i' values differ too much from the X_i; but if the variations are small, one may estimate new Y' with some confidence of getting at least the right direction of change.

Given the framework above, Blackett went on to suggest a variety of "statistical" and "theoretical" methods, as well as mixtures of the two, for estimating the derivatives of Y with respect to the X_i. Many of these methods involve what might now be termed simple combat models.

The *presentation* stage is, of course, the destination that the operations research worker must keep in mind throughout his work in all the other stages. The presentation is not a report to be neatly filed but rather a recommendation for action. The recommendation may go to a high-level decision maker, but sometimes it is most effective when it is communicated to operators directly (as, e.g., when it concerns tactics under their control). Morse (in Koopman et al. (13)) points out that, on occasion, the recommended "action" may be to retain present procedures unchanged.

Careful thought should be given to the presentation of recommendations to executives. Blackett (1) cautions that the operations research worker:

> ...should only recommend to the executives that the action should be taken if he himself is convinced that he would take the action, were he the executive authority. It is useless to bother a busy executive with a learned resume of all possible courses of action and with the conclusion that it is not possible to decide between them. Silence here is better than academic doubt.

Morse and Kimball (19) note that the different functions of scientist and administrator may produce some tension. The scientist by profession is skeptical, whereas the administrator must make decisions. If to the scientist it seems that some decision is unwarranted or arbitrary, he should "not consider that this is necessarily a repudiation of his work."

In reflecting on all of the five stages discussed above, it is important to recall that their successful accomplishment in wartime depended, to a considerable extent, on the working mode of the operations research analysts. Morse (in Koopman et al. (13)) recalls:

> We did live with our problems and with our clients in a way that I don't think happens often nowadays. We worked pretty much seven days a week, 52 weeks a year. It wasn't a matter of a three-week or a three-month study and leaving a report which gets filed. We had to live with our successes as well as our failures.

COMBAT DATA

As seen above, the acquisition of data in the observation phase is an essential component of the methodology of combat operations research. It is interesting to note that there is also a "non-real-time" use: in recent years there has been much talk of using historical combat data in testing large computer models. On the one hand, combat data are potentially of value because they come from a real war. On the other hand, however, the quantity of high-quality data is very limited.

It is not easy to get good data in war. We often lack data about our own operations because our own troops are typically too busy or preoccupied with combat to be concerned with data recording. To an even greater extent, we may lack good data about enemy operations because an enemy takes special precautions to conceal his movements and may actively seek to mislead us.

It may be noted, therefore, that in many combat operations research problems it took a great deal of effort to acquire the data needed to estimate the numerical values of the "Y" and "X_i" essential to Blackett's variational approach. Given the effort and ingenuity required, operations research practitioners hardly viewed combat models as "data-rich." The value of such data in improving operations was typically so great, however, sometimes even without much in the way of formal models, that efforts to acquire good data were deemed highly worthwhile.

Analysts soon learned to be skeptical of data routinely collected. A typical experience is illustrated by Blackett (2). It was noted during World War II that coastal antiaircraft batteries in Britain were apparently shooting down twice as many night bombers per shell fired as inland batteries. All kinds of explanations were suggested until:

> ...suddenly the true explanation flashed into mind. Over the land a battery, say, in the Birmingham area, which claimed an enemy aircraft destroyed, would have this claim checked by the search for and identification of the crashed machine. If no machine were found, the claim was disallowed. But, in general, no such check could be made of the claims of the coastal batteries, since an aircraft coming down at sea would not be traceable. The explanation of the apparent better shooting of the coastal batteries turned out to be due...to an overestimate...of their claims...by a factor of about two.

Jones (12) gives a similar example where RAF fighter pilots claimed 186 killed but German records inspected after the war showed that the German losses were 62 aircraft. Jones observed that "it was quite unreasonable to expect any pilot who thought that he had destroyed an enemy aircraft to follow it down...and watch it crash...." Hence overclaiming was inevitable.

Dorfman (6) mentions the difficulty of estimating ranges and angles from a moving aircraft:

> Vision is extremely unreliable, and optical illusions are frequent. In part this is due to... the decreased supply of oxygen at altitude...to a lack of reference points and the scale of the distance involved.... In part it is due to the inability of the eye to make allowance for the fact that it is on a platform which moves at better than 300 feet per second....

Sometimes the realization of the low quality of certain kinds of data came only when an analyst got out into the field. Jones (12) recalls an example where two operational research workers had the problem of deducing the characteristics of German magnetic mines laid at sea, "especially the sensitivity and polarity of the firing mechanism." The data from which the characteristics were to be deduced were the reports of our minesweepers as they exploded the mines, with positions of the explosions being reported as ranges and bearing from the minesweepers. One of the operational research workers took a few trips on a minesweeper to sample the data for himself and immediately found that "the reports from the minesweeping crews were wildly inaccurate as regards both range and bearing, and the only item of data on which one could rely was whether the mine had exploded to port or starboard." Interpreting the data with this limitation in mind, he did solve the problem, whereas the other researcher stuck to his desk and never reached an answer.

Once it was realized that supplementary sources of data were needed, much more use of photography was made. Fighter interceptors carried "gun cameras" and bombers carried "bomb cameras." Miser (in Koopman et al. (13)) recalled that in the 8th Air Force in Britain in World War II:

> ...the analysts discovered that...there was a contingent of public relations people, and a few airplanes were carrying cameras for the purpose of taking pictures to be sent back to the *New York Times* and other...publications. These were magnificent cameras; they took contact prints about 12 inches square...the analysts...with a little bit of practice, discovered that you could do an incredible amount of analysis from these photographs.... So the next job was to see if you could get the General...to put cameras like that in at least one

> airplane in every formation. Then you had to persuade the military that the analyst needed a copy of those photographs as much as the *New York Times*. Well, that took a bit of doing, but they got it done....

Such data sources were much used in World War II. In Korea and Vietnam they were again introduced. During the Korean conflict the analysis of gun camera film was accompanied by the analysis of supplemental combat information reports designed to give information that could not be obtained from film. In instituting such reporting systems, it is highly desirable for an analyst to explain in person why the data are important and how they should be obtained.

By the time of Vietnam, there were many reporting systems, some in the form of "all-purpose" computerized data banks. These were not as useful as had been hoped. In reporting on an Army Operations Research Symposium of 1969, which was entirely devoted to problems of data collection, Greene (9) commented:

> 'Data banks' are difficult to use for operations research, etc., because they are not designed with specific problems or analyses in mind. We were admonished to first plan an analysis, decide what to do with data, and then plan and design the data collection.

The frequently rough character of combat data also related to the "hemibel thinking" described by Morse and Kimball (19). Their principle contains two important ideas. The first, and most frequently cited, is that in early World War II successful applications of operations research often resulted in improvements by factors of at least three (a "hemibel" is a factor of the square root of 10, slightly more than 3, whereas on the same "logarithmic" scale a "bel" is a factor of 10). The other idea underlying "hemibel thinking" is the one that derives from the character of combat data. It was found that unless there was a ratio of three or more between a theoretical value and an observed value, or between two different observed values, one might very well suspect that the difference was not significant. This may not seem to be a sophisticated statistical test, but it was a very useful rule of thumb in screening out many data that warranted little further analysis.

Incidentally, it was the second (data-related) idea of "hemibel thinking" that proved to be more fundamental for combat analysis. That is, it was always wise to view data with some skepticism, and the hemibel rule helped to screen them. As time went by, the other role of hemibel thinking (in screening research projects), however, became less relevant. For example, Solandt (24) recalled of World War II:

> By the time I came into the Army Operational Research Group, the spectacular successes had already been made and it was a question of plugging away at the steady improvement of operations in many fields, older fields, where a great deal of work had already been done to perfect organizations and systems.

DATA AND MODELS IN COMBAT ANALYSIS -- EXAMPLES

Analysis is, of course, the pivotal step in the five-stage process of problem perception-formulation-data collection-analysis-presentation. It is by analysis that one decides whether to accept the hypothesis conjectured in the formulation stage, and press on towards recommendations for presentation to decision makers, or whether the hypothesis should be rejected or refined, more data collected, etc. If the hypothesis involved an explicit model, then that model will come to the fore during analysis. This section will give a few examples of simple models, and a few examples where no explicit model seems to enter.

The examples come from World War II, Korea, and Vietnam (more generally, Southeast Asia to include neighbors of Vietnam). The focus is on the more "real-time" analysis which most deserves to be called wartime combat operations research. The emphasis, that is, must be on wartime analysis that is done in time to have impact while the war is still going on. Each of the three combat situations considered here gave rise to much valuable work completed after the combat had ended. Some of this work was technique development (like the previous examples of linear programming and network flow algorithms), some was related to model development, some to model verification (or the estimation of important parameters), and some to the provision of context for weapons system research and development. Often such work is more "publishable" and more theoretically "elegant" than combat analysis itself, and sometimes it colors our impressions of what combat analysis is like.

Actually, considering the very different geographical extent of World War II, Korea, and Vietnam (and their different settings in time), combat analysis in the three conflicts had remarkably similar character. World War II, with its far-flung theaters of operation, provided examples of more different kinds of combat than either Korea or Vietnam. Operations research in World War II was newer, with less of a menu of developed techniques, and in that sense was more challenging. The urgency of research perhaps seemed greater also, to the extent that World War II was seen as an "all-out" war, whereas Korea and Vietnam were conflicts conducted in "far-off" Asia while we were "at peace" in the rest of the world. Yet the analysis itself, in each of the three situations, exhibited many of the same themes and much the same character.

Silber (23), for example, in a Korean study (for the 5th Air Force of the United States Air Force) that computed force requirements where bomber aircraft are to attack targets of different sizes, noted:

> ...that our assumptions and methods are similar to those used by many of the research organizations working on statistical and probability problems for the Services during World War II.

An interesting parallel between World War II and Vietnam is shown by two following observations. Summing up experience in Air Force analysis in World War II, Leach (16) wrote:

> The staple grist in all air forces was bombing accuracy, bomb selection and fuzing, gunnery and radar.

In Vietnam, as combat began to heat up in 1962, analysts picked up very much the same kind of problem. In a covering letter to an "after action" report of analysis (Starkey (25)), it was noted that operations analysis and intelligence analysts:

> ...assumed the responsibility for weapon selection and physical vulnerability, and by recommending the introduction of different air ordnance into South Vietnam and matching the ordnance to the various target situations, increased the effectiveness of both VNAF and USAF air operations by some factor conservatively exceeding two.

When analysts first attacked bombing problems in World War II, they inherited no tradition of combat operations research or body of established technique other than the "scientific outlook" itself. As noted in previous sections, they introduced as an MOE the percent of bombs falling within 1,000 feet of a designated impact point, and they derived numerical values of the MOE by analyzing photographs taken by "bomb cameras." These were powerful developments and led to great improvements. Leach (16) recalls a case where the bombing data by themselves were conclusive:

> With a workable tool in their hands, the analysts began getting some startling results. Some group commanders directed each bombardier to sight independently for range, i.e., along track (there could be only one sighting for deflection, i.e., across track, since the bombers had to maintain formation and thus keep in a predetermined position with reference to the lead ship). Others ordered all bombardiers to drop when they saw the bombs leave the lead ship. Still others dropped on the squadron leaders. There were strong *a priori* arguments vigorously pressed for the validity of each of the three techniques. A comparison was now possible. Here is the result:

Technique of Sighting	Bombs within 1000 ft. of Aiming Point
On Group Leader	24.0%
On Squadron Leader	11.8%
Independent Sighting for Range	8.3%

> That ended the argument! Sighting on the group leader became standard operating procedure.

In the previous example one had what was essentially an "unplanned experiment." Three different procedures were being tried, and the data, once available, led directly to a recommendation to select one of the three without further experimentation or modeling (though data were used continuously, of course, to monitor accuracy and to provide the basis of other improvements). Leach (16) gives a slightly more complicated example where data are again critical but no mathematical model was involved. In this example, however, the data led to rejection of an initial hypothesis and the need for a revised hypothesis before a recommendation was made:

> When the Eighth Air Force was small, it was confidently predicted that larger formations would saturate enemy defenses and that the accuracy of bombing would rise sharply. The first groups, it was argued, would absorb the fighter attack and the fresh vigor of the flak; then later groups would have a field day. But when five or more groups could be sent over the same target, bombing was worse than before. An OAS analysis of these strikes with references to the other in which each group went over the target told the story:

Order over Target	Bombs within 1000 ft. of Aiming Point
1st	23.4%
2nd	16.3%
3rd	9.9%
4th	7.8%
5th and over	4.2%

The data show a difference in effectiveness between the first and last group of a factor exceeding five (clearly exceeding Morse and Kimball's "hemibel" of "three plus") and lead one to reject the hypothesis. Additional thought led to an explanation that appealed to two factors expressed by Leach as:

> 1) congestion of air traffic over the target and consequent interference with bombing runs,
>
> 2) obscuration of the target by bomb falls from preceding groups.

Leach reports no model development but observes that:

> Conferences between OAS and command personnel brought a decision to space out combat wings over the target at intervals of 1 to 5 minutes. The

results were shown in the following table, the payoff appearing in the last column:

Order over Target	Bombs within 1000 ft. of Aiming Point		
	Before Change	After Change	Improvement
1st	23.4%	26.5%	13%
2nd	16.3%	19.8%	21%
3rd	9.9%	22.2%	124%
4th	7.8%	20.1%	158%
5th and over	4.2%	19.8%	370%

The 13% improvement for the first group was attributed to a general increase in effectiveness of bombing throughout the command, but the bulk of the improvement shown by the later groups seemed to be attributable to the new tactics.

It would be misleading to leave the impression that all improvements in bombing accuracy came with such seeming ease. The account of Brothers (4) indicates that many factors had to be considered and intricate calculations were involved. The effect of the analysis, which represented the work of many operations research scientists, was to increase the percent of bombs within 1,000 feet of the aiming point from about 15 in 1942 to about 60 in 1944. The point of the examples above is to show that data, which are essential for analysis and the scientific use of models, may in themselves point the way to dramatic improvements in effectiveness.

Because data were seen to be very important, both in themselves and as a prelude to more extensive analysis, compilations of data tend to be published periodically during wartime by a variety of offices. This was true in both Korea and Vietnam. The use of data and statistical analysis was very widespread in the analysis of the Vietnam conflict, and it is interesting to note two extreme examples, one from the Office of the Secretary of Defense and the other from the "field."

During the 1960s, systems analysis became an essential tool of the new planning, programming, and budgeting system. It was institutionalized in the Office of the Assistant Secretary of Defense (Systems Analysis), and its products were actively used by the Secretary of Defense as a basis for many decisions. As the Secretary became intimately involved in the conduct of the war in Southeast Asia, which saw much more centralized direction from "Washington" than previous conflicts, it was natural that the Systems Analysis office would play an important role.

One of the products of the office was a monthly publication of an analysis report (over several years including the late 1960s and early 1970s). Its purpose was stated (ODASD (21)) as follows:

> *The Southeast Asia Analysis Report* is not an official publication of the Department of Defense, and does not necessarily reflect the views of the Secretary of Defense, Assistant Secretary of Defense (Systems Analysis), or comparable officials. It is published by the Office of the Deputy Assistant Secretary of Defense (Systems Analysis), Regional Programs, as a stimulant to further analysis and discussions. The intent of the report is to update and disseminate, for information and comments, studies of interest to those in OSD and the Services who are responsible for decision making or analysis related to the conflict in Southeast Asia. We publish about the 20th of each month.
>
> We are trying to improve the quality of analysis on SEA problems and to increase the audience for interesting analytical work. We do our best to ensure accuracy of data and interpretation, but we reserve the right to say next month that what we said this month is wrong. We will print rebuttals and reanalyses of previously published reports. We draw heavily on data from the OSD SEA Statistical Summary published by OASD (Comp), Statistical Services. We will not duplicate the content of that document -- which is the only official source of SEA statistics in OSD.

As one might suspect from the mention of statistics in the statement above, the character of the analysis was quite similar to that of the World War II bombing problems discussed earlier. The acquisition of data and its "first-order" analysis constituted most of the quantitative analysis, with little elaborate modeling. Table 1 indicates the kind of problem addressed, as illustrated by a sampling of 13 of the SEA analysis reports.

The entries in Table 1 indicate some of the many concerns that lay outside the scope of traditional military operations. There was deep concern with the establishment of a viable government in South Vietnam, and such threats as terrorism and economic instability were more worrisome than large concentrations of opposing troops. Rather than

Table 1. Topics from Southeast Asia Analysis Report

TITLE OR SUBJECT	APRIL 1968	JULY 1968	OCTOBER 1968	MARCH 1969	APRIL 1969	MAY 1969	JULY 1969	OCTOBER 1969	JANUARY 1970	MARCH 1970	JUN/JUL 1970	SEP/OCT 1970	JAN/FEB 1971
DEPLOYMENTS	●		●										
FRIENDLY FORCES	●	●	●	●	●	●	●						
REVOLUTIONARY DEVELOPMENT	●												
AIR OPERATIONS	●	●	●	●		●	●				●		
VIET CONG/NV ARMY		●	●										
PACIFICATION		●				●				●		●	
ECONOMICS		●							●				
CASUALTIES					●	●	●	●	●			●	●
REDEVELOPMENTS							●	●	●				
ANTI-INFRASTRUCTURE CAMPAIGN								●					
REPUB VN AIR FORCE								●	●	●			
INDICATORS OF ENEMY ACTIVITY									●				
ALLIED OPERATIONS									●				
NVA/VC PRISONERS OF WAR										●			
HAMLET EVALUATION SYSTEM											●		●
TERRORISM IN SVN											●	●	
PHOENIX												●	
THE WAR IN CAMBODIA													●
WHAT THE VIET PEASANT THINKS													●
STANDOFF ATTACKS													●
THE HERBICIDE ISSUE													●

DATE OF REPORT

fighting at a well-delineated front against masses of enemy soldiers, we were fighting many "local" actions and often saw the most challenging battle as that "for the hearts and minds" of Vietnamese villagers.

For this kind of struggle there existed fewer combat models. A typical article in the SEA analysis reports made its case with presentation of data -- indicators and MOEs -- intended to show trends over time, comparisons of different troops, etc. Most of this kind of analysis involved neither sophisticated statistics nor elaborate model construction. There was, however, some regression analysis, both simple and complicated in terms of the number of variables involved. An interesting simple example is a model devised by Safeer of the Weapons Systems Evaluation Group (reported in TIA (26), Headquarters US Air Force) to predict enemy-initiated attacks in South Vietnam. It represented the number of such attacks in a given month as a linear function of two variables -- the total number of trucks sighted in Laos one and two months earlier and the total number of trucks sighted in a particular area of North Vietnam two and three months earlier.

At the other extreme, as opposed to the "centralized" use of analysis in "Washington," there was also much use of combat analysis in the "field" during the Vietnam conflict. Very interesting examples are given by Ewell and Hunt (7) of analyses conducted during 1968, 1969, and 1970 to improve operations of the 9th Infantry Division and II Field Force. Some indication of the diversity of applications is given by their chapter headings, which refer to unit management and personnel actions, Army aviation assets, battlefield analysis, intelligence, tactical refinements, pacification, etc.

The subtitle of the above is "The Use of Analysis to Reinforce Military Judgment," and this indicates the character of the applications. The authors observe:

> Dr. Enthoven has stated in...*How Much Is Enough?* that operations research was used very little in Vietnam. He is probably correct if he was referring to systems analysis and cost-effectiveness.... The Vietnamese war was so complex, so varied, and so changeable that these high-level approaches tended to be very difficult to use with any confidence. However, simple straightforward operations analysis, while not easy to use,...was used extensively.

"Straightforward operations analysis," as reported by Ewell and Hunt, made extensive use of data collection and presentation to indicate to commanders how operations were going. It was difficult to construct models of combat processes in Vietnam, the authors found, and they emphasize the virtues of flexibility. As we introduced improved tactics, the enemy reacted quickly so that after a few months further improvements were needed to retain the initiative. Computers were useful, but a "general problem was the rigidity of computer data bases" because these tended to become outdated.

So far, the examples and discussion of this section have sought to convey the extremely important role in wartime combat analysis of ingenious data collection and presentation. It has also been indicated that statistical procedures such as regression analysis may lead to models that are useful for understanding basic relationships, prediction, etc. Since the theory of probability provides so much of the underpinning of the statistical analysis of data, the previously asserted importance of probability might already be inferred.

The role of probability is still broader, however, than the foregoing may suggest. In this connection it is useful to apply what Koopman (14) said of the role of probability in the theory of search, *mutatis mutandis*, to combat operations research generally:

> Every operation involved in search is beset with uncertainties; it can be understood quantitatively only in terms of the concepts and laws of the scientific theory of probability. This may now be regarded as a truism, but it seems to have taken the developments in operational research of World War II to drive home its practical implications. The reason for the lateness in taking full advantage of this theory has undoubtedly been the simple nature of most of the earlier operations of search, the short ranges involved, and the power of the human eye at such ranges. These factors made a 'deterministic' approximation satisfactory: 'When a thing is there and I look at it without anything in front of it, I see it; otherwise I don't.' Approximations of this sort are still useful, but they must be validated by more realistic analysis, based on probability, among other things.

For a variety of such reasons, therefore, many useful combat operations research models were largely applications of probability, and an introduction to probability theory was considered essential in any systematic account of operations research methodology. Thus Morse and Kimball (19) devoted an early chapter to probability, which introduced, for example, binomial, normal, and Poisson distributions. A later chapter applied probability to "Gunnery and Bombardment Problems." The latter, incidentally, discussed a "sampling" approach that today would be termed "Monte Carlo."

Because the outcome of combat is "beset with uncertainties," one might expect every combat model to have a strong probabilistic component. In many models, however, the role of chance variation may not be very evident. This can happen, for example, when it is adequate to use expected values, or, in general, when some "law of large numbers" applies so as to suppress the impact of variation.

It is important to note also that applications of probability in combat operations research range over a very wide spectrum from the simple to the extremely subtle. In his much quoted pioneering paper, Blackett (2) spoke simply of "a grasp of fluctuation phenomena (i.e., Poisson's Distribution)" as a requirement for an operational research practitioner. Similarly, Solandt (24) recalled that:

> ...we rarely used anything more elegant than some simple statistics and probability. I think our most spectacularly successful mathematical tool was the knowledge that in adding errors you do not add them arithmetically but add the squares and take the square root. This knowledge led to many important discoveries.

On the other hand, Newman (20) in characterizing the methods of early operations research workers observed:

> Their indispensable tool was the mathematics of probability and they made use of its subtlest theories and most powerful techniques.

Certainly such developments as the theory of search in World War II, which is reported in an account by one of the pioneers that is an operations research classic

(Koopman (14)), included some of the most subtle and powerful applications.

In some problems of wartime combat operations research one uses a model, or an *additional* model, because it supplements a previous approach. As often happens in mathematics or science, the availability of two or more ways of solving a problem greatly deepens our insight and adds to our confidence in the solution.

As an illustration, it is interesting to compare, in abbreviated versions, two approaches to the analysis of encounters between convoys of allied merchant vessels and packs of German submarines in World War II.

The first approach, reported by Morse and Kimball (19), gave some interesting results that were derived by statistical analysis of data on encounters between convoys and submarines. Letting

M = the number of merchant vessels in a convoy,
K = the number of merchant vessels sunk,
C = the number of escort vessels protecting the convoy,
N = the number of submarines in a pack, and
L = the number of submarines sunk,

analysis of the data led to

$$K = 5N/C \tag{1}$$

and

$$L = NC/100 \tag{2}$$

as good approximations. From these expressions, one can derive the exchange ratio

$$L/K = C^2/500 \tag{3}$$

Of even greater interest, however, was the observation that equation (1) is independent of M, so that the loss *rate* varies inversely with M. Contrary to a belief that the British Admiralty had held, it appeared that large convoys were safer than small convoys! Since operations analysis of the data had led to a result both unexpected and significant in its operational implications, it was natural to ask if the result could be corroborated by other approaches.

One such approach, reported by Blackett (2), made use of a simple model. This model expressed the probability of sinking a merchant vessel in a convoy as a product of three probabilities:

$$P_r(\text{sinking}) = \underset{(a)}{P_r(\text{convoy sighted})} \times \underset{(b)}{P_r(\text{penetration given sighting})} \times \underset{(c)}{P_r(\text{sinking given penetration})} \quad (4)$$

Both theory and data suggested that the probability in (a) was independent of convoy size. It depended rather on the relationship of submarine search practices and convoy routes, speeds, etc. Simple models suggested that the probability in (b) depended on the linear density of escorts screening the convoy but not on the number of merchant vessels in the convoy. To obtain the probability in (c) it was noted that once a submarine had penetrated, its expected number of sinkings was independent of convoy size; thus the probability in (c) that a particular merchant vessel would be sunk varied inversely with convoy size. When the three probabilities on the right-hand side were multiplied together, the same inverse relationship extended to the overall P_r(sinking).

There is an interesting postscript to the above account that has valuable lessons for operations research generally. Despite the statistical analysis and the plausibility of the results from the simple model, Blackett did not immediately urge the Admiralty to reverse its long-standing preference for small convoys. Only after deep soul-searching, in which he became convinced that his family would be safer in a large convoy were they to sail to Canada, did he report the result and point to its striking implications.

During the period of Vietnam analysis such "dual" approaches continued to be used, now with the possibility of more complicated models executed on electronic computers. One illustrative example (Freck (8)) reported some analysis by the Weapons Systems Evaluation Group (WSEG) of combat aircraft losses in Southeast Asia. The two approaches were described as follows:

> Data from many sources (including standard reporting systems, studies, and information collected by WSEG representation in the theater) are used in an attempt to relate changes in each factor to changes in losses. Where the quantitative data

> from SEA are insufficient for analysis, simple theoretical models and studies based on SEA experience and using typical SEA inputs are utilized.
>
> The second method of analysis consists of a relatively detailed mathematical model (simulation) of the engagement between maneuvering aircraft and antiaircraft artillery.... While the model was developed in part to be useful in explaining past experience in SEA, it could also be used to determine effects of changes in offensive/defensive environments, both in SEA and elsewhere.

As a footnote to the above reference to a reporting system, there was the following observation:

> Earlier WSEG reports have commented upon the data base problem as pertains to accuracy and completeness as well as the unavailability of data for some study purposes. This problem continues to exist and has had its impact upon this study although considered effort was made to circumvent these data shortcomings.

While the above example indicates that the "dual" approach was being used in Vietnam as it had been in World War II, it also illustrates some differences. The computer model was developed in part with an eye to the future (the "war to end all wars" had become "ancient" history), and the data problems in Vietnam constrained the usefulness of large models.

An example of an application of a computer model that was more directly aimed at Southeast Asia operations as the primary target is given by Bortels and English (3). This example concerned the effect of aircraft formation size and configuration on the effect of their jammers in countering enemy radars. The computer was used, in effect, to speed the solution of the kind of geometry problem that in World War II would have required a more tedious approach.

There was another kind of model application, emphasized by Morse and Kimball (19), that was introduced in World War II operations research, perhaps as a transplant of scientific and engineering practice. This was the comparison of an observed value of an operational parameter with a theoretical value of the parameter calculated for ideal conditions. An example given by Morse and Kimball involves "sweep rates" calculated for aircraft, surface vessels, or submarines engaged in searching ("sweeping out") an area thought to contain enemy vessels or submarines. Operations research studies compared "observed" values of Q_{op} (the "operational sweep rate") to "ideal" values of Q_{th} (the "theoretical sweep rate").

The theoretical sweep rate was given by

$$Q_{th} = 2RV \quad (\text{sq. mi./hr.}) \tag{5}$$

where

R = effective lateral range of detection in mi.
V = average speed of searching craft in mi./hr.

If one searched with sweep rate Q_{th} for time T, then the expected number of contacts would be

$$C_{th} = Q_{th}\ T\ (N/A) \tag{6}$$

where N is the number of enemy craft in area A, so that N/A represents an average density. Similarly, the observed number of contacts, C_{op}, would be given by

$$C_{op} = Q_{op}\ T\ (N/A) \tag{7}$$

which gives

$$Q_{op} = (C_{op}/T)(A/N) \tag{8}$$

From intelligence information on enemy operations, reasonably good values of A/N could often be inferred, so that from observed values of C_{op}, values of the "operational sweep rate" could be calculated via (8). By comparing these operational values with the theoretical values of (5), operations research workers derived useful indications of operational effectiveness.

In some cases, (8) assumed a very simple form. When the density N/A was thought to be essentially constant, (8) was proportional to the first part of the expression

$$C_{op}/T \tag{9}$$

which is just the observed contact rate. Increases in this ratio would flag improvements in weapons or tactics, etc., so that it often became quite a useful MOE.

Through the process illustrated above -- comparison of observed operations with idealized models -- and through the solution of many individual problems that pointed to common patterns, there came the recognition of some basic relationships and models as quite general and fundamental. As such models were introduced, and related operational constants were measured, there developed the understanding of some operations that had many aspects of a scientific theory. Thus, in World War II operations research there came to be a "theory of search," a "theory of bombardment," etc.

As such bodies of knowledge were established, operations research workers gained confidence in using some models to estimate requirements, to indicate preferred allocations of effort, and to prescribe tactics. An example of a very simple relation that is found in several models is

$$P = 1 - \exp(-kE) \tag{10}$$

where a measure of effectiveness P is expressed in terms of input effort, E, and the effectiveness of a unit effort, k. This expression applied, for example, in search theory where P denoted the probability of finding an object, E denoted the time of random search, and k depended on the effective sweep rate and target density.

The very similar expression

$$P = 1 - (1 - p)^E \tag{11}$$

was used in gunnery (and bombardment) problems where P denoted the probability of destroying a target by expending E shells (or bombs) and p denoted a single weapon probability of destruction. One notes that (11) may be put in the form of (10) by defining

$$k = -\ln(1 - p) \tag{12}$$

Expressions (10) and (11) apply where the effort E is all devoted to finding or destroying a single target. If the effort were applied to a multiplicity of targets, then the expected value of the MOE (contacts in the case of search and targets destroyed in the case of gunnery or bombardment) would be a linear function of the effort (see, for example, (7)). Sometimes, however, the military situation dictates concentration on a single target, and (10) applies. One may note that k itself could be viewed as an MOE at a lower level, and an expression of the form

(10) shows how, in some operations, it "translates into" an MOE at a higher level.

As such basic expressions became the cornerstones of developing theories, they were sometimes used as the point of departure for more involved mathematical developments. An excellent example of what can result is Koopman (14). This contains chapters on "The General Operations of Search"; "Position, Motion, and Random Encounters"; "Target Detection -- Individual and Cumulative"; "Target Localization"; "Time Evolution of Target Positions"; "The Optimum Distribution of Searching Effort"; "Search in Time: Information and Jeopardy"; "The Practical Search for Targets in Transit"; "Sonar Screens: Old Examples and New Developments"; and "Aerial Escort During World War II." Some of its results were incorporated in official Navy publications in World War II.

It is important to note that in keeping with the "scientific outlook" the validity of the applications of basic expressions, their extensions, and the body of theory generally must all be the subject of continuing scrutiny. Even when the functional forms introduced for one combat situation apply in another, the operational "constants" may have to be modified. The skeptical outlook is especially important when models derived from combat analysis are appropriated as building blocks for large computer simulation models. Because of the very size of the latter, it is often difficult to determine the full impact of some component after it has been embedded. Hence, when one takes a successful combat model to be the core of a subroutine of a large computer model, he should exercise the same caution that was appropriate to wartime combat analysis.

Finally, it is interesting to inquire into the role of the famous Lanchester equations in wartime combat operations research. Lanchester (15) introduced the equations initially as a possible means of investigating the impact that the airplane, then only recently invented, might have on the nature of war in the future. Thus, at the beginning of World War II, Lanchester equations and other differential equations of a similar nature were known to some of the scientists who became active in operations research.

Blackett (2) saw in any effort to use such equations an example of what he termed an *a priori* method that, though it might yield interesting theoretical results, was not nearly as practical as the "commonsense" variational approach. Morse and Kimball (19) reported several extensions of Lanchester's original equations that were investigated during World War II and gave as their evaluation:

> As usual, the constants of warfare are not very constant, and only approximate forecasts can be obtained. In most cases such constants are not known sufficiently accurately to warrant their being used in mathematical equations of any complexity.
>
> Occasionally, however, it is useful to insert these constants into differential equations to see what would happen in the long run if conditions were to remain the same as far as the constants go. These differential equations, in order to be soluble, will have to represent extremely simplified forms of warfare; and therefore their range of applicability will be small.... Nevertheless, it sometimes happens that considerable insight can be obtained into interrelationships between measures of effectiveness by studying differential equations involving them.

In the years since those words were written, the introduction of electronic computers and convenient programming languages has made it possible to solve very complicated differential equations numerically. In fact, some of the models discussed elsewhere in this monograph may be viewed as difference equation versions of gigantic sets of Lanchester-like equations. This interest in building large models had begun by the time of Vietnam, but it appears that there was little impact of such Lanchester-like large models on combat analysis at that time. Data problems alone would probably have been enough to inhibit much use.

There was one kind of application, much in the spirit of Lanchester's original work, that was made during the Vietnam conflict. This was the application of Lanchester-like equations to gain insight into the nature of guerrilla warfare. Deitchman (5) used a very interesting variation on Lanchester's original equations. Lanchester had discussed two basic cases: one where both sides have "aimed" fire (which gives the famous "square law") and one where both sides have "area" fire (which gives the "linear law"). Deitchman modeled ambushes by letting the ambushers have "aimed" fire and letting the ambushees have "area" fire (which gives a "mixed" law). Schaffer (22) later extended this approach, introducing more detail and distinguishing conditions of skirmish, ambush, and siege.

So far, it would appear that the chief product of the use of Lanchester-like equations, especially those simple enough "to be soluble" without computers, is insight. Since operational "constants" are often not very constant or very available, there are obvious constraints on the use of such equations. Insight, however, is itself a valuable commodity in the problem-solving process, as in problem formulation, and the use of simple models -- like Lanchester equations -- can have the same kind of value that "thought experiments" have in physics.

A LOOK AHEAD

Given the examples of the past, it is very tempting to take a look ahead. Such a look can be *predictive* -- what is likely to happen -- or *prescriptive* -- what ought we try to make happen. A simple projection of past experience with combat analysis would show a great danger of forgetting valuable lessons learned in trying circumstances, and it would suggest trying to plan in peacetime to be better prepared for future combat analysis should the need arise.

There are current developments in technology that may lead to marked changes in the future conduct of combat analysis, and incidentally, some fusing of the predictive and prescriptive approaches. For over ten years there have been predictions of the instrumented battlefield and a proliferation of sensors. Similarly, there have been predictions of still greater advances in computer speed and memory size than we have yet seen. The use of computers of all sizes has already made great changes in weapons and military management systems.

In World War II, Korea, and Southeast Asia, there was strong emphasis on the collection and analysis of data bearing on the performance of guns, bombs, radars, etc. In each conflict, it was necessary to develop, essentially from scratch, the procedures and organizations required. One is led to predict that, in any future combat, far more of these required capabilities will be already built-in. More and more of our procedures are being automated and computerized; analysis "packages" are increasingly widespread.

As these new capabilities become available, they will make possible great improvements in combat analysis. Much of the drudgery of data collection and tabulation can be circumvented. Some of this improvement may come without much deliberate planning. We can do much better than this, however, if we carefully identify some of the analysis needs that are common to modern conflicts and prepare to meet them.

SUMMARY

The main examples of important combat analysis in wartime have come from World War II and, to a lesser extent, from conflicts in Korea and Vietnam. These examples suggest that such analysis, including the most successful results, owed remarkably little to sophisticated models, either mathematical or computer. This remained true even in the Vietnam conflict, though military operations research and computer simulation had by that time made rather sophisticated techniques readily available.

The operations research method, as pioneered in World War II, depended strongly on the most fundamental aspects of the "scientific outlook." Essential to the successes were skepticism, energy and ingenuity in digging for "facts," deep thought and imagination in discerning the purpose of an operation and how to measure effectiveness, and persistence and caution in seeking operational improvements. These activities, as in scientific investigation generally, often led to model building, but the models tended to be much simpler than some of the very large computer models developed in the last decade or two.

With a "war on," there is less incentive for, and less interest in, the investment of great effort in building a massive computer model to "substitute" for the war. On the one hand, the war itself serves as an "overall model," a collection of "test ranges," etc. On the other hand, the complexity of a war; the variety of operations; the quality of the data (real but often "dirty," partly as a result of enemy concealment and "disinformation"); and the obvious importance of leadership, morale, courage, and luck all serve to remind us vividly of how hard it is to model combat on the grand scale.

The examples of wartime combat analysis serve also to remind us of aspects of the scientific approach that we sometimes forget as we build large models untested in war.

Always in order are a healthy skepticism, energy and ingenuity in digging for insights into the nature of combat, and deep thought and imagination in forecasting the operational environments in which our forces might be employed. Finally, as we build larger models, it is important to remind ourselves of the supplementary value of simple models and operations research approaches as used so effectively in past combat analysis.

REFERENCES

1. *P.M.S. Blackett, "The Scope of Operational Research," Operational Quarterly, vol. 1, no. 1, p. 6, March 1950.*

2. *P.M.S. Blackett, Studies of War, Hill and Wang, New York, 1962.*

3. *W.H. Bortels and J.A. English, Computer Program to Aid in Evaluation of ECM Formations, Hq. PACAF Operations Analysis Note 68-5, May 1968.*

4. *L.A. Brothers, "Operations Analysis in the United States Air Force," Journal of the Operations Research Society of America, vol. 2, no. 1, pp. 1-16, February 1954.*

5. *S.J. Deitchman, "A Lanchester Model of Guerrilla Warfare," Operations Research, vol. 10, no. 6, pp. 818-827, November 1962.*

6. *Robert Dorfman, Work and Method of Operations Analysis, Hq. U.S. Air Force, Directorate of Operations, Operations Analysis Division Working Paper 17.1.2, July 1951.*

7. *LT GEN J.J. Ewell and MAJ GEN I.A. Hunt, Jr., Sharpening the Combat Edge, Dept. of the Army, Washington, DC, 1974.*

8. *Peter G. Freck, Analysis of Combat Aircraft Losses in Southeast Asia, Weapons Systems Evaluation Group Report 128, April 1968.*

9. *William G. Greene, Army Operations Research Symposium -- 1969, Hq. U.S. Air Force Operations Analysis, AFGOAC Memorandum for Record, June 1969.*

10. Charles J. Hitch, "Suboptimization in Operations Research Problems," Journal of the Operations Research Society of America, vol. 1, no. 3, pp. 87-99, May 1953.

11. Charles J. Hitch, "An Appreciation of Systems Analysis," Journal of the Operations Research Society of America, vol. 3, no. 4, pp. 466-481, November 1955.

12. R.V. Jones, The Wizard War -- A Story of British Scientific Intelligence, 1939-1945.

13. B.O. Koopman et al., Panel Discussion ("World War II Operations Research," unedited transcript), Operations Research Society of America meeting, Los Angeles, 13 November 1978.

14. B.O. Koopman, Search and Screening, Pergamon Press, New York, 1980.

15. F.W. Lanchester, Aircraft in Warfare: The Dawn of the Fourth Arm, Constable & Co., London, 1916.

16. COL W. Barton Leach, U.S. Army Air Forces, Professors of Air Combat, Hq. U.S. Army, Operations Analysis Section, 1945.

17. Robert Leckie, The Wars of America, Harper and Row, New York, 1981.

18. MITRE, National Security Issues Symposium -- 1981, MITRE Document M82-30, 1982.

19. Philip M. Morse and George E. Kimball, Method of Operations Research, OEG Report No. 54, Operations Evaluation Group, Office of the Chief of Naval Operations, Navy Department, Washington, DC, 1946.

20. James R. Newman, The World of Mathematics, Simon and Schuster, New York, 1956.

21. ODASD (Office of the Deputy Assistant Secretary of Defense [Systems Analysis], Regional Programs), Southeast Asia Analysis Report, March 31, 1969.

22. M.B. Schaffer, "Lanchester Models of Guerrilla Engagements," Journal of the Operations Research Society of America, vol. 16, no. 3, pp. 457-488, May 1968.

23. Jack Silber, Force Requirements for Fighter Bombers, Hq. Fifth Air Force, Operations Analysis Office, Memorandum 59, January 1953.

24. Omond Solandt, "Observation, Experiment, and Measurement in Operations Research," Journal of the Operations Research Society of America, vol. 3, no. 1, pp. 1-14, February 1955.

25. Lawrence Starkey, End-of-Tour Report, Hq. Second ADVON Operations Analysis, June 1962 (with transmittal letter by Dr. Oldham, Hq. U.S. Air Force Operations Analysis, AFCOA, July 1962).

26. TIA (Trends, Indicators, and Analyses), Hq. U.S. Air Force, DCS Plans and Operations, Directorate of Operations, October 1968.

27. W.F. Whitmore, "Edison and Operations Research," Journal of the Operations Research Society of America, vol. 1, no. 2, pp. 83-85, February 1953.

28. W.F. Whitmore, "Military Operations Research -- A Personal Retrospect," Operations Research, vol. 9, no. 2, pp. 258-265, March-April 1961.

CLAYTON J. THOMAS

Mr. Thomas has been developing and applying methodology for Air Force studies since 1947: first at the Institute for Air Weapons Research, University of Chicago (1947-55); then in the Air Force Office of Operations Analysis (1955-71); and in the Office of the ACS, Studies and Analyses, HQ USAF (1971-present). From 1946 through 1949 he studied and taught mathematics at the University of Chicago. During World War II he served as a weather officer in the US Army Air Force. His lectures on military operations research have been translated into several languages for use as a text. A member of Phi Beta Kappa, he has received the ORSA Lanchester Prize, the Air Force Exceptional Civilian Service Award, and the Air Force Association Citation of Honor. Mr. Thomas has been active in ORSA (Council Member, Associate Editor Operations Research, and Chairman of the Committee of Ethics and Professional Practice) and in MORS (a past President and currently Air Force Sponsor's Representative).

ABOUT CHAPTER 2

If one is going to evaluate a model it helps to know its specific ancestors, their virtues and vices. Next best is a *general* history of the application of models. In this chapter, J.J. Martin traces the development of models and model applications for nuclear warfare since the earliest days.

In the 1960s and 1970s, Martin says, there was an overfocus on computer simulations of nuclear warfare, especially simulations of nuclear exchanges, to the exclusion of other analytic forms. The results were an excessively narrow range of scenarios; neglect of many important operational realities; an emphasis on nuclear firepower and neglect of maneuver, tactics, command and control, and operational timelines in nuclear warfare; and the analysis of nuclear forces in isolation from other military forces and from the political-military context of general-purpose conflict.

Computer models that came into vogue in the 1960s inhibited strategic thought by focusing DoD staffs on what they could calculate with these models -- stylized nuclear weapon exchanges -- rather than on the outcomes of military campaigns. Strategy has moved from simple nuclear exchanges to incorporate comprehensive problems of nuclear warfare. But the capabilities of staffs and their supporting contractors have lagged behind that movement.

To date, models of nuclear warfare have been of greatest value in defining nuclear warfare issues, identifying force posture and C^3I deficiencies, and providing technical support on important details. Further, they impose a measure of discipline on debates about nuclear planning issues by keeping these debates rooted in physical and operational realities. Such models, however, have been less helpful in treating the larger nuclear policy, employment planning, and force acquisition issues confronting national security decision makers and staffs.

Martin concludes with concrete guidelines for improving nuclear warfare modeling and analysis.

Modeling Nuclear Warfare
by
J.J. Martin

> When we consider that...research [into wars of the future] proposes to discover the nature of the cataclysm which may come upon humanity, and that analysis of it cannot be accomplished except by exercise of the imagination within the confines of rigid logic, it becomes a fascinating study.
>
> Giulio Douhet
> *The Probable Aspects of the War of the Future* (1928)

One of the more esoteric problems confronting the military operations analyst is modeling of nuclear warfare. Except for the two atomic bombs used by the United States against Japan in 1945, mankind has no experience with such a conflict; in fact, the circumstances, delivery systems, and targets associated with the employment of atomic weapons in 1945 were such as to render even this experience of virtually no value in modeling nuclear warfare, except as a source of nuclear effects data. Yet the enormous destructive consequences of a nuclear war, the ways that the threat of such a war influences strategy and foreign policy, the large sums spent annually be the United States on strategic and theater nuclear forces -- all this and more make the systematic modeling of nuclear warfare a matter of continuing importance.

Early in the atomic age, analysts started to develop and apply models of nuclear warfare, with mixed success in helping those charged with formulating nuclear policy and making decisions about the US nuclear force posture.† The following paragraphs trace the evolution of nuclear warfare modeling, examine how key problems of modeling nuclear warfare have been approached, survey the current state of these modeling efforts, and assess where models of nuclear warfare have helped and where they have been dysfunctional. This assessment finds that models have provided helpful technical support on many details in the analysis of nuclear warfare, but that they have not been nearly as helpful in dealing with the larger policy, employment planning, and force acquisition problems confronting national security decision makers, even though modeling has considerable potential in these areas. Indeed, use of a specific style of nuclear exchange computer modeling that came into vogue in the 1960s has inhibited realistic approaches to these difficult policy and planning problems.

By *nuclear warfare* we mean any conflict in which nuclear weapons are used against military or civilian targets. This definition intentionally is broad, and is meant to include:

- General war -- a worldwide war between the United States and the Soviet Union, involving other nations as well, in which at some point nuclear weapons are used. General war analyses should focus on both general-purpose and nuclear force operations prior to, during, and after large-scale homeland-to-homeland nuclear exchanges.

- Limited nuclear war -- a conflict in which nuclear weapons are used, but in which the objectives of the various sides, the geographic scope of the conflict, and the level of violence are more limited than in general war. The risks, consequences, and dynamics of escalation to general war should be an important part of limited nuclear war analyses, as should be military outcomes if escalation is kept under control.

---

†*In 1948, P.M.S. Blackett analyzed strategy, force requirements, and arms control issues in the atomic age, using scenarios, threat projections, effects data, calculations of force interactions, and other elements of what today would be called a "model"; see his* *__Military and Political Consequences of Atomic Energy__*, *Turnstile Press, London, 1948.*

- Superpower crises -- by implication, the analysis of nuclear warfare potentially includes the analysis of crisis dynamics, military alerting, indications and warning, command and control systems and information management, and other aspects of deep crises among the nuclear powers, in which the prospects for nuclear war are an ominous backdrop.

NUCLEAR WARFARE ISSUES AND APPLICATIONS OF MODELS

The abiding nuclear warfare issues for the United States and its allies are what deters the Soviet Union from taking a variety of hostile actions, what are the proper roles for nuclear weapons, how should military operations be conducted in a nuclear war, what nuclear forces should the United States and its allies have, and what nuclear arms control approaches should be pursued. These issues relate to a wide range of military forces, including strategic offensive and defensive forces; offensive and defensive theater nuclear weapons deployed with ground, air, and naval forces; roles, characteristics, and employment of general-purpose forces in an integrated nuclear, chemical, and conventional combat environment; and the supporting command, control, communications, intelligence (C^3I) and target acquisition systems. The specific issues addressed in nuclear warfare analyses are themselves wide-ranging, including issues in the following categories:

- Objectives, strategy, and missions for nuclear forces.

- Research, development, and procurement of nuclear weapons, their delivery systems, and the supporting C^3I systems.

- The size, mix, and deployments of nuclear forces and the stockpile of nuclear weapons.

- Planning for the employment of nuclear forces, including concepts of operations, doctrine, and tactics; development of specific targeting plans; transition to nuclear warfare; and the conduct of integrated conventional, chemical, and nuclear operations.

- Threat projection and assessment.

- Adequacy of C^3I systems and procedures in nuclear war.

- Civil preparedness for nuclear war.

- The nature of future conflict milieus.

- Net assessments of the military capabilities of two or more adversaries.

The Overview of this monograph defines seven different applications of models -- battle planning and analysis, analysis during wartime operations, selection among alternative weapon system candidates, determination of the overall size and mix of a military force, planning for utilization of human resources, planning for logistics support to military operations, and assisting the development of national policy. Models of nuclear warfare have been developed and applied almost entirely for the purposes of battle analysis, weapon selection, determination of force level and mix, and national policy analysis. Models have not, of course, been developed during actual nuclear warfare for analysis of the ongoing operations, as was done for conventional force operations in World War II. Human resource models have included personnel planning to support nuclear weapon systems, but the models themselves have been influenced (probably rightly so) by the nature of human resource planning, not by the characteristics of nuclear warfare. There has been virtually no modeling of logistics support in the context of nuclear warfare. This is a neglected area to which recent analysis of protracted nuclear war scenarios has been directing attention. Several aspects of logistics require consideration in the modeling of nuclear warfare, including logistics support to strategic and theater nuclear forces held at high readiness levels during a prolonged crisis or conventional conflict; the performance of logistics support to general-purpose forces in an integrated conventional-chemical-nuclear combat environment; and reloads for strategic and theater nuclear weapon delivery systems in a general war.

The relative weight of emphasis on battle models, models for weapon system selection, and models for force sizing purposes has varied over time. The nature and purposes of nuclear warfare models have, quite naturally, followed the evolution of nuclear strategy and the specific issues under examination.

The 1950s: Emerging Conceptual Framework. During the 1950s, the strategy of the United States and NATO was to use threats of massive nuclear retaliation to deter a wide range of Soviet aggressive actions. The prevailing image of warfare was that of a general war, nuclear almost from the outset. This was a period in which the United States and Soviet Union were building up both intercontinental and theater nuclear forces. Nuclear forces were seen as an affordable substitute for the general-purpose forces the West did not believe it could deploy in sufficient numbers to hold their own against Soviet general-purpose forces.

This was also a period of intense questioning of the official strategy, both inside and outside the Government. The result was the emergence of a richer and more complex framework for thinking about nuclear warfare that strongly influenced the strategy and forces of succeeding decades. Concepts of first strike and second strike; alternate scenarios for nuclear warfare; limited war concepts; the importance of warning and graduated responses to ambiguous warning; trade-offs among offensive and defensive forces, including civil defenses; improved understanding of the subtleties of deterrence -- these were the products of the 1950s.

Under the influence of these inquiries, development and application of models followed many exploratory lines in the 1950s. The early nuclear war studies were large-scale, very detailed campaign analyses that were limited by the capabilities of computers in that period. These efforts proved too costly, and a number of different approaches were taken to reduce the costs and improve the responsiveness of models of nuclear warfare.† These approaches included the following:

- Use of relatively simple models of operational processes, integrated into a sophisticated analytic framework for examination of specific problems such as measures to improve the survivability of strategic air bases.

---

†For a summary of strategic war gaming developments in the 1950s, see Garry D. Brewer and Martin Shubik, The War Game: A Critique of Military Problem Solving, Harvard University Press, Cambridge, 1979, pp. 59-66.

- Applications of gaming to strategic force and air defense acquisition problems, with detailed "map-gaming" support.†

- Development of Monte Carlo computer simulation techniques for detailed analysis of specific operational processes.

- Development of political-military gaming techniques for analysis of the political and strategic aspects of nuclear crises and conflicts.

- Advances in the mathematical theory of games.

- Development of techniques, such as Delphi, for systematic use of human judgment in analyses.

The 1960s: Strategic Exchange Modeling. During the 1960s, the United States and NATO shifted strategy from massive retaliation to a combination of conventional force emphasis and threats of escalation to nuclear war (called "flexible response") to deter aggression in Europe. At the intercontinental level the changes in strategy were not as abrupt. As technologies contributing to ballistic missile counterforce, air defense, and anti-ballistic missile defense capabilities began to emerge in the 1960s, the strategic objective of limiting damage to the United States received greater emphasis. On the other hand, perceptions about the cost-effectiveness of various strategic objectives also led in the 1960s to strong emphasis, especially in strategic force acquisition, on the capability to ride out a Soviet first strike and inflict "unacceptable" retaliatory damage on Soviet cities. This doctrine came to be known as mutual assured destruction, as the Soviet buildup of their own secure retaliatory force of strategic delivery systems increasingly was seen to hold US cities at risk. The doctrinal, force acquisition, and employment policy debates that ensued from these

---

†As used here, the term gaming implies (1) competition between two or more sides, (2) participation by humans on at least one side (e.g., to represent decision makers), and (3) an output that is distinct from the game play itself. The output or analytically usable results from a game may, for example, consist of a strategy, a campaign outcome, or improved understanding of what factors are most important in a complex military operational problem.

not entirely consistent strategic concepts focused on such topics as offensive counterforce capabilities; ballistic missile defense; air defense; civil defense; prelaunch survivability and defense penetration effectiveness of US strategic forces; connectivity of strategic communications systems sufficient to execute the US retaliatory attack after a large Soviet nuclear strike; roles for theater nuclear forces in a flexible response strategy for the European theater; and alternative approaches to strategic arms limitation.†

In support of the analysis of these issues, the Department of Defense invested substantial sums in the development and application of quantitative models of strategic nuclear exchanges. Use of map games and political-military games for analysis declined (although games were still used for training and educational purposes in the 1960s), and complex computer models moved more strongly onto center stage. This was the period of development of elaborate strategic exchange models using mathematical allocation algorithms (e.g., Code 50, Arsenal Exchange Model). Models of prelaunch survivability, urban-industrial damage, and the functioning of air defense and ballistic missile defense systems became more sophisticated and elaborate. During the latter half of the 1960s, more effort was placed on special-purpose computer models to analyze the characteristics of such advanced weapon concepts as MIRVs and cruise missiles. Modeling of theater warfare generally focused on conventional force operations, with analysis of theater nuclear force issues being carried out off-line from computer models of conventional conflict. There was, however, some growth during the 1960s in computer models of collateral damage to civilian targets resulting from the use of nuclear weapons.

<u>The 1970s: Modeling Lags Behind Changing Strategic Concepts</u>. In the 1970s, there was greater awareness of Soviet concepts of warfare, which differ markedly from Western concepts in many respects. Concerns about the prelaunch survivability of US strategic forces grew more prominent in this period as Soviet strategic counterforce

---

†*For an overview of the strategic issues and analytic approaches of the 1960s, see Alain C. Enthoven and K. Wayne Smith, <u>How Much Is Enough? Shaping the Defense Program, 1961-1969</u>, Harper and Row, New York, 1971.*

capabilities continued to improve. US military strategy moved away from the polarized debates of the 1960s about damage-limiting versus mutual assured destruction to a more subtle, and equally controversial, emphasis on deterrence through improved warfighting capabilities, as well as through threats of extensive destruction of the political-economic structure of the Soviet Union. Other key elements of US military strategy in the 1970s were concepts for a broad range of nuclear attack options, including those for limited nuclear war, and development of a common conceptual framework for dealing with both strategic and theater nuclear force issues.

Strategy and analysis in the 1970s began to focus more heavily on such issues as the future viability of the triad of strategic forces (ICBMs, SLBMs, and manned bombers); the limited use of nuclear forces and the control of escalation; modernization of theater nuclear forces; planning for combined conventional-nuclear operations in the European theater (concerns about military operations in a chemical warfare environment began to emerge late in the decade); recovery after a large nuclear war; support to the SALT and MBFR negotiations; and net assessments of the strategic and theater nuclear force balances.†

The relationship between strategy and models of nuclear warfare underwent an important change in the 1970s. During the 1960s, model development adapted rapidly to the analytic requirements generated by the emerging strategic concepts of damage-limiting and mutual assured destruction; indeed, the evolution of US nuclear strategy in the 1960s was in part conditioned by models of strategic exchanges in a Department of Defense increasingly dominated by quantitative support to programmatic decisions. In contrast, there has been considerable

---

†Many of the key strategic issues of the 1970s are discussed in a collection of the more important unclassified papers of the decade in Robert J. Pranger and Roger P. Labrie, eds., Nuclear Strategy and National Security: Points of View, American Enterprise Institute for Public Policy Research, Washington, 1977. For a more general overview of strategic issues from the late 1940s through the 1970s, see Laurence Martin, ed., Strategic Thought in the Nuclear Age, The Johns Hopkins University Press, Baltimore, 1979.

lag between the evolution of strategic concepts in the 1970s and the development of models for use by DoD staffs and contractors to support analysis of issues emerging from these strategic concepts.

This situation has come about for several reasons. In part, it is due to the normal delays encountered as changes in strategic thought filter down from the top levels of the Department of Defense to the organizations and staffs that ultimately must implement these changes. More importantly, however, in the 1970s there were fewer organizational and intellectual ties than in earlier periods between those responsible for changing directions in strategic concepts and those concerned with model development and application. In the 1970s, changes in strategic concepts came about in spite of the prevalent computer models of nuclear warfare, not as a result of insights produced by these models. In fact, one of the factors responsible for strategic shifts in the 1970s was the dissatisfaction of the post-McNamara DoD leadership with the inability of the prevalent models to address many aspects of nuclear warfare that were increasingly recognized to be of importance in deterring, controlling, or terminating a nuclear war with the Soviet Union. The differing issues and analytic demands generated by the strategic concepts of the 1970s (as compared with those of the 1960s) are illustrated in Table 2, which shows differences in strategic issues that result, depending on whether the emphasis is on strategic nuclear exchanges or on warfare in which nuclear weapons are a major, but not the exclusive, type of force involved.

Another factor responsible for the considerable lag between the changes in strategic concepts of the 1970s and the development of models to support the analysis generated by these changes is institutional in nature. The far-reaching transformation wrought by Robert McNamara in the approach of the Department of Defense to the force acquisition process embedded quantitative analysis and computer models deep in the bureaucratic structures of the Pentagon. Consequently, DoD staffs found themselves dealing in the 1970s with a more complex set of strategic employment concepts within a planning, programming, and budgeting system still strongly oriented to narrowly focused quantitative support to force acquisition decisions. The pace of day-to-day staff work did not allow the time required to replace old models with new ones; the staffs had little choice in the near term but to continue

Table 2. Strategic Issues Resulting from Different Emphases in Nuclear Strategy

ELEMENTS OF STRATEGY, PLANNING, AND ANALYSIS	EMPHASIS OF STRATEGY	
	NUCLEAR EXCHANGES	NUCLEAR WARFARE*
1. DETERRENT THREATS	WIDESPREAD DESTRUCTION	DENIAL OF ENEMY MILITARY OBJECTIVES
2. NATURE OF WARTIME OBJECTIVES	NUCLEAR RETALIATION	POLITICAL-MILITARY OBJECTIVES SPECIFIC TO THE CONFLICT; ESCALATION CONTROL; WAR TERMINATION
3. FORCE SURVIVABILITY	PRELAUNCH SURVIVABILITY UNDER SUDDEN NUCLEAR ATTACK; PENETRATION OF ENEMY DEFENSES	ENDURANCE; SURVIVABILITY UNDER NON-NUCLEAR ATTACKS; RECONSTITUTION; RESIDUAL NUCLEAR BALANCES
4. COMMAND, CONTROL, AND COMMUNICATIONS	COMMUNICATIONS CONNECTIVITY TO LAUNCH RETALIATORY STRIKE	ADAPTIVITY AND CONTROL OF MILITARY OPERATIONS
5. FORCES OF INTEREST	INTERCONTINENTAL FORCES; STRATEGIC DEFENSES	THEATER NUCLEAR FORCES; GENERAL-PURPOSE GROUND, AIR, AND NAVAL FORCES
6. FORCE EMPLOYMENT PLANS	LARGE, PREPLANNED NUCLEAR STRIKE OPTIONS	RANGE OF FLEXIBLE NUCLEAR ATTACK OPTIONS; RETARGETING DURING WARTIME; COMBINED CONVENTIONAL-CHEMICAL-NUCLEAR OPERATIONS
7. CONCERNS OF PEACETIME DIPLOMACY	ARMS CONTROL FORMULAS	WAR AVOIDANCE; CRISIS MANAGEMENT APPARATUS

* THESE ISSUES ARE ADDITIVE TO THOSE RESULTING FROM AN EMPHASIS ON NUCLEAR EXCHANGES.

to use the old models in order to meet the deadlines of the planning, programming, and budgeting system.

This situation was exacerbated by the difficulties of adapting large-scale, exchange-oriented computer models to the more complex problems of nuclear warfare illustrated by Table 2. Such attempts were made, but they were not very satisfactory because the basic structure of computer models of strategic nuclear exchanges is not readily adaptable to capture the major additional dimensions in the modeling of nuclear warfare required by the strategic concepts of the 1970s. Thus, short-term evolution of strategic exchange models of the 1960s made little headway in dealing with the strategic issues emerging during the 1970s. The longer-term approach of developing new models of nuclear warfare also made slow progress in the 1970s because of the inherent difficulties of quantifying many of the emerging concepts and issues. DoD staffs and contractors, laboring under 15 years of cultural conditioning to strictly quantitative support to decision making, were slow to appreciate that a richer approach to modeling nuclear warfare was needed, one that artfully combined a variety of qualitative, quantitative, and judgmental elements, more in the manner of the strategic analysis of the 1950s.†

A final factor inhibiting the development and application of improved models of nuclear warfare in the 1970s was the organizational separation between staffs and contractors concerned with modeling strategic nuclear warfare and those concerned with modeling the operations of general-purpose forces. Some of the modeling techniques needed for analysis of nuclear warfare issues were

†In 1978, a Defense Science Board study reviewed DoD capabilities to assess the strategic balance, identified a number of major problems with current methods of strategic force analysis, and recommended the development of a war-gaming approach to overcome these problems. In 1979, the Department of Defense initiated a program to develop an improved war-gaming style of strategic analysis to evaluate strategic forces in a wide range of scenarios; to consider strategic force capabilities in the broader context of military operations; to allow inclusion of a richer set of operational factors; and to integrate such traditionally neglected topics as space warfare, ASW, and C^3I into strategic force analyses.

available in general-purpose force models, but communications between the two communities were sparse and haphazard, in part because the strategic concepts of the 1960s had not encouraged this communication. Transference of general-purpose force modeling techniques to strategic analysis has been slow and only began to take place in the late 1970s.

Thus, modeling of nuclear warfare proceeded along two lines in the 1970s. One line of development continued the evolution of nuclear war computer models of the 1960s, with more sophistication and complexity to handle specific force acquisition or targeting issues that persisted into the 1970s. There were, for example, more detailed models of prelaunch survivability and penetration of US strategic forces. These included models of Soviet SLBM attacks on US strategic bomber bases, models of ballistic missile defense of US ICBMs, and the Air Force's Advanced Penetration Model of the attrition inflicted by Soviet air defenses on mixed forces of manned bombers and cruise missiles. Another example of this line of development is the evolution of more elaborate models of strategic communications systems to measure connectivity and message delays, taking into account physical damage to the systems, nuclear radiation and electromagnetic effects, and the effects of enemy jamming. Yet another example is the evolution of more detailed models of collateral damage from the employment of strategic or theater nuclear forces. Trends in this area in the 1970s were toward more detailed models (e.g., for fallout patterns); integration of prompt and delayed collateral effects into a single model; and the modeling of secondary effects on cities, such as the spread of fires resulting from nuclear detonations.

The second line of model development in the 1970s sought in more fundamental ways to deal with the emerging strategic concepts of nuclear warfare. Political-military games were used to explore the utility of various types of limited nuclear options. Some effort was devoted to modeling the economic recovery of the Soviet Union after a large nuclear attack, in support of ongoing strategic force targeting studies. The most prominent effort of this sort, however, was the work that began in the latter half of the 1970s to model the use of theater nuclear forces in ground campaigns. Many of these efforts modeled use of nuclear weapons by only one side. The Institute for Defense Analyses (IDA) developed its TACWAR model of ground combat in the late 1970s, which includes provisions for two-sided use of nuclear and chemical weapons. More recent efforts include ongoing work to incorporate theater nuclear and chemical use into other theater-level models, and development of a two-sided ground campaign model that seeks to capture the effects of delay and disruption (as well as of attrition) resulting from the use of theater nuclear, chemical, and conventional weapons.

The 1980s: Modeling Nuclear Warfare. The start of the 1980s is marked by uncertain movement within the United States and NATO toward even more emphasis on nuclear warfighting and toward major programs of strategic and theater nuclear force modernization. The strategic concepts and issues of the early 1980s reflect a continuation of the thinking of the 1970s and include such topics as targeting of Soviet political and military controls; protracted general war; endurance and reconstitution of nuclear forces and C^3I; operations on an integrated conventional-chemical-nuclear battlefield; Navy theater nuclear forces; the US nuclear posture in the Asia-Pacific area; and theater nuclear force arms control. Model development and applications in the early 1980s in the area of nuclear warfare continue the trends of the 1970s, with slow movement toward more complete treatment of the evolving issues of nuclear warfare. The modeling techniques of the early 1980s are still heavily oriented toward computer models, but there is increasing use of other techniques, including net assessment of trends and asymmetries in nuclear forces; scenario narratives; tactical and strategic gaming using computer support; interactive man-machine computer models; and applications of board game techniques to the modeling of nuclear warfare.

PROBLEMS IN MODELING NUCLEAR WARFARE

A key difference between nuclear and conventional warfare is the relative lack of historical experience and operational data related to nuclear war. This difference conditions the "modelability" of nuclear warfare by constraining the types of models that can be developed credibly and the uses to which these models can be put. For example, some uncertainties about nuclear war cannot be resolved short of actually having such a war, and analysts must work with models in ways that are consistent with these special, unresolvable uncertainties.

These observations underscore a related point -- modeling of nuclear warfare must be preceded by systematic efforts to understand the basic phenomena involved and the ways in which nuclear war may be similar to or different from other more familiar forms of warfare. There is a tendency to take the approach of "Let's model it so we can understand it," rather than the more difficult but more sophisticated approach of "Let's understand it so we can model it."†

Much of the work on strategic concepts since the early 1970s has been directed at better understanding of what a nuclear war might be like. While many uncertainties remain, enough is understood about potential nuclear wars to reveal a number of problems with many contemporary models of nuclear war, problems that are only slowly being corrected. Occasionally, these modeling problems are exacerbated by organizational parochialism or even poor analysis, but largely they exist because of the reasons discussed above -- the lack of organizational and intellectual ties between those who are driving the emerging strategic concepts and those who are developing models of nuclear warfare, the difficulty of making progress on some of these modeling problems within the structure of large-scale computer models of strategic exchanges, and the lack of sustained interaction between modelers of strategic force operations and modelers of general-purpose military operations.

The following modeling problems, while not exhaustive, are certainly among the most important:

- Excessive modeling of US or NATO use of nuclear weapons without considering nuclear responses by the Soviet Union.

- Modeling of nuclear force employment in isolation from other key elements of military operations.

- Failure to model important operational factors and uncertainties.

- Modeling the results of nuclear weapons employment in terms of casualties and damage rather than in terms of the impact on mission accomplishment.

---

†I am indebted to Milton Weiner for this trenchant formulation of a classic problem.

- Modeling too narrow a set of nuclear war scenarios.

- Modeling nuclear warfare from US perspectives, and not those of the Soviet Union or US allies.

These recurrent problems reflect the way that Western defense establishments think about nuclear war, a way of thinking that increasingly is being seen as inadequate in light of the Soviet approach to these matters, the problems posed by long-term political-military competition between the United States and the Soviet Union, and the prospects in the next decade or two for proliferation of nuclear weapons. Thus, discussion of these modeling problems is instructive in its own right and leads below to a set of criteria by which to assess models of nuclear warfare in the context of problems to which they are likely to be applied in the future.

<u>One-Sided Nuclear Use</u>. Modeling the one-sided use of nuclear weapons without considering nuclear responses -- or nuclear preemption -- by the other side was an early problem with strategic force analyses. By the 1960s this problem was corrected through development of prelaunch survivability models and strategic exchange models in a doctrinal context that increasingly emphasized second-strike strategic force capabilities. More recent use of two-sided games for strategic force analysis and imaginative development of scenarios that reflect Soviet concepts of operations (such as preemption of US nuclear use) reinforce the practice of modeling two-sided nuclear use in strategic force analyses.

Unfortunately, this is still a problem with many models and analyses of theater nuclear force employment, which all too often model the results of NATO employment of nuclear weapons without including Warsaw Pact nuclear responses. The result, of course, is overly optimistic assessments of the net military consequences of US or NATO first use of nuclear weapons. One reason this problem persists in theater nuclear force models is that already complex models of ground campaigns become even more complex, expensive to develop, and time-consuming to use if two-sided nuclear use is incorporated. A more fundamental reason is that many users of analyses still do not appreciate the integral role of nuclear weapons in Soviet military doctrine, and have not demanded of modelers the tools for analyses of two-sided theater nuclear use. In other words, the doctrinal context for theater nuclear force analyses still does not require routine consideration of two-sided nuclear use to the same degree as does the doctrinal context for strategic force analyses.

While there are elements of nuclear warfare analyses in which it may be appropriate to model only one-sided use (e.g., prelaunch survivability analyses of strategic or theater nuclear delivery systems and C^3I), these generally will be the minority of cases. It is the obligation of model developers to keep the reality of Soviet or Warsaw Pact nuclear responses (and nuclear preemption) before the users of analyses, and to develop and apply models that facilitate routine inclusion of enemy nuclear responses when an analysis of US or NATO first use is undertaken.

Modeling of Nuclear Forces in Isolation. A second recurring problem with nuclear warfare models is that they frequently treat nuclear forces in isolation from other key aspects of military operations such as the political-military objectives, strategy, and doctrine of each side; the general-purpose force operations that would be in progress at the time of nuclear use; and the command and control of combined conventional-chemical-nuclear operations. The result is incomplete analyses of nuclear force capabilities, analyses that do not consider such issues as: the effect of enemy surprise and deception on US or NATO capabilities to carry out nuclear operations and to exploit such operations with general-purpose forces; the attrition of nuclear forces by conventional or chemical attacks; the selection of appropriate and feasible targets for nuclear forces; the impact of nuclear weapons employment on the pace of the general-purpose conflict; the timing of nuclear use vis-à-vis conventional operations; the synergistic assessment of combined nuclear and general-purpose force operations; the trade-offs between nuclear and conventional capabilities; and a whole host of other issues.

Limited progress in dealing with this problem is being made through the upgrading of theater campaign models to include nuclear and chemical use, as discussed above (e.g., the Army's model improvement program, IDA's TACWAR model). Gaming has the potential to set the use of strategic or theater nuclear forces into a broader political-military context, but few applications of gaming for analytic purposes have been made to date. Moreover, gaming is not the appropriate form of model for every analysis and, even when it is appropriate, the costs and setup times currently associated with gaming inhibit its applications.† The best approach currently available to the problem of excessive analysis of nuclear forces in isolation from other aspects of military operations is to carry out the analysis using a variety of tools (narrative analysis, scenario design, simple mathematical models, computer models for selected analyses, etc.), with the logic of the issue under analysis guiding both the selection of aspects of the military operational context to be modeled in detail and the final integration of the elements of the analysis into an assessment of nuclear capabilities in the broader operational context.

A particularly important deficiency in models of nuclear warfare is command and control. Connectivity analyses of communications systems are routinely performed as part of the evaluation of strategic communications alternatives, but such analyses are rarely carried over into analyses of nuclear delivery system capabilities. Information uncertainties about launch failures, damage to enemy targets from earlier strikes, empty enemy ICBM silos, and the like are sometimes reflected in force allocation models. But the specific C^3I and target acquisition requirements of individual strategic or theater nuclear forces are seldom examined; survivability of political or military commanders, realistic consideration of information available in wartime for direction of combined conventional-chemical-nuclear operations, time delays in information for retargeting, and many more C^3I matters are rarely considered in nuclear warfare analyses. No significant progress has been made to date toward integrating C^3I considerations into nuclear warfare models. This should be a matter of high priority for research on model improvements.

Neglect of Important Operational Realities. Even in models and analyses that seek to examine nuclear forces in the broader context of general-purpose force operations, important operational realities and uncertainties are all

---

†*With suitable computer and data base support to a gaming center, game setup time can be compressed significantly and the manpower associated with gaming reduced. This would, however, require a substantial one-time investment in a computer-supported gaming center.*

too frequently ignored by the models. The list of these neglected operational factors is long. Examples are:

- Operational limitations on target acquisition systems supporting the use of theater nuclear forces against movable targets.

- Delays, errors (sometimes gross errors), and gaps in information available to military commanders and political authorities directing nuclear force operations.

- Uncertainties that affect the operational capabilities of many nuclear weapon systems (e.g., uncertainties about target hardness, uncertainties about the performance of ICBMs and SLBMs that have not been tested under real-world operational conditions).

- A full range of enemy threats and tactics. For example, models rarely include ASW campaigns against SSBNs or unconventional warfare operations against theater nuclear forces.

- Movement of a large number of civilians on the roads of Eastern and Western Europe during combat operations. Civilian traffic attempting to flee the combat zone could seriously impede the dispersal of NATO's theater nuclear forces; this traffic is rarely taken into account when estimating collateral damage.

Inclusion of such operational factors in models of nuclear warfare generally has not been demanded by the users of analyses, even though some factors could be included in available models without undue difficulty. There is, however, growing recognition of this problem, and model developers should exercise initiative -- first to include neglected factors that are easy to handle and, second, to carry out research on techniques to include the many important operational factors that are by no means trivial to model effectively and efficiently. Some progress has been made in models of specific combat processes. For example, computer models of SLBM attacks on bomber bases have become rather sophisticated in their inclusion of the relevant operational factors. As another example, some computer models of theater nuclear force survivability are now including Soviet agents and long-range reconnaissance patrols, enemy correlation of multiple sources of information, and decision algorithms to reflect enemy command and control processes.

Generally, however, models of nuclear warfare do not give sufficient attention to many important operational factors. This statement applies to gaming as well as to computer models. For the most part, recent efforts to include neglected operational factors in nuclear warfare analyses have moved away from quantitative models, using such means as historical or scenario narratives, narrative illustrations of the analytic consequences of the failure to consider important operational realities, and net assessments of trends and asymmetries related qualitatively to a wide range of operational considerations. But these qualitative treatments are usually not integrated well -- if at all -- with the outputs of quantitative models to produce a complete analysis.

Inadequate Modeling of Consequences of Nuclear Weapons Use. A related problem is the treatment of the consequences of nuclear weapons employment in models of nuclear warfare. There are two issues. The first is the handling of nuclear effects data. There are important uncertainties in available nuclear effects data; examples include the effects of varying levels of radiation on the performance of combat personnel, the performance of specific pieces of electronics equipment in a nuclear environment, and the collateral casualties produced by nuclear detonations. Uncertainties arise for different reasons, many of which cannot be reduced prior to a nuclear war. For the effects of radiation on troop performance, the uncertainties stem from a sparse data base; for performance of electronics equipment, they are due to variations in the physical configuration of specific pieces of equipment; for collateral casualties, the uncertainties result from variations in shielding provided by structures in specific towns and from such situation-specific variations as the sheltering posture of civilians, the distribution of people in and around the towns, and the direction and intensity of the prevailing winds at the time of nuclear detonations.

Considerable progress has been made in the development of data bases, models, and computer codes to quantify the uncertainty in the damage caused by nuclear detonations, but these uncertainty assessments are generally not included in the prevailing models of military operations in a nuclear environment. In fact, military combat models are not generally set up to facilitate determination of whether nuclear effects uncertainties are important for a particular analysis.

The second, perhaps more fundamental, issue connected with the consequences of nuclear weapons employment is the difficulty with current models and data bases of translating casualties and damage into an evaluation of the ability of military units to carry out their missions. Models of ground force campaigns tend to treat firepower more realistically than maneuver, yet many of the more important consequences of nuclear weapons employment have to do with delay and disruption, which impact on the movement and coordination of military units. More generally, most models do not handle reconstitution efforts very well; there is poor understanding of how much damage a military unit such as a battalion or division can sustain and still function as a coherent operational force, and there are virtually no data relating to the psychological impact of nuclear attacks on combat troops. These problems are becoming more prominent as the Army has begun study of operations on an integrated conventional-chemical-nuclear battlefield, which has highlighted difficulties in evaluating integrated effects on military units because of differences in the basic structure of data bases. Conventional munitions effects data generally relate varying levels of conventional munitions employment to the probability of mission impairment for the unit under attack.† While nuclear effects data for fixed targets are structured in terms of mission impairment, nuclear (and chemical) effects data for combat personnel are structured in terms of the physical functions of individuals, not in terms of the probability of overall mission impairment for the unit composed of these individuals.

Some progress is being made in treating the problem of translating casualties and equipment damage from nuclear detonations into measures of unit combat capability. Notable are recent Army efforts to model the combined effects of delay, disruption, casualties, and equipment. But more work is needed on data base structures, modeling techniques, and the integration of these results with

†*There are, of course, many uncertainties in conventional effects data that are not customarily reflected in models and analyses of general-purpose conflict. For a critical review of these problems, see J.A. Stockfish, Models, Data, and War: A Critique of the Study of Conventional Forces, The Rand Corporation, Report R-1526-PR, Santa Monica, CA, March 1975.*

other combat model outputs before modelers of nuclear warfare can claim to handle adequately the problem of translating the results of nuclear weapons employment into overall effects on military operations.

Narrow Set of Scenarios. Another recurring problem is that available quantitative models of nuclear warfare treat too narrow a set of scenarios and, in fact, constrain analysts from considering a wider range of scenarios because of the assumptions built into the basic structure of prevailing computer models. The result is that these models fail to consider important dimensions of nuclear warfare because of scenario limitations. These neglected scenario dimensions include, for example:

- Prolonged periods of superpower crises in which the endurance of nuclear forces and the airborne elements of C^3I systems on advanced alert became important.

- Protracted periods of conventional war prior to nuclear force employment. Endurance of nuclear forces on high alert during conventional conflict and attrition of nuclear forces by enemy non-nuclear attacks are among the important factors that tend to be overlooked if models of nuclear warfare do not examine periods of conventional conflict.

- Protracted periods of nuclear war in which force reconstitution, C^3I endurance and reconstitution, and nuclear reloads may be important.

- Space warfare during any part of the conflict spectrum, which may degrade or deny tactical warning of enemy nuclear attacks and substantially reduce communications capacity.

Existing computer models of nuclear warfare do not readily adapt to diverse scenarios that go well beyond the strategic exchange structure of these models. Available models of general-purpose conflict have the potential to treat some of the neglected scenario dimensions discussed above (e.g., ASW campaigns against SSBNs), but such applications are rarely made. Some recent specialized models of theater nuclear force prelaunch survivability treat conventional and unconventional force attacks. Generally, however, diversity in nuclear warfare scenarios has been handled through narrative illustrations, narrative scenario development efforts, and gaming. The best approach available today to the problem of an excessively narrow set of nuclear warfare scenarios is to structure an analysis using logic and judgment, supported at appropriate points in the scenario by discrete application of quantitative models.

Excessive Modeling From US Perspectives. US models of nuclear warfare overwhelmingly reflect US perspectives, not those of the Soviet Union or of US allies. Yet Soviet perspectives are crucial for analyses that purport to assess the deterrent or escalation control properties of alternative nuclear force postures or targeting doctrines. Allied perspectives are important for assessing the adequacy of US nuclear forces to support allies or for examining the political-military implications of nuclear force balances.

During the 1970s, substantial progress was made in understanding Soviet concepts of warfare and how they differ from Western concepts. Much of this material, including a substantial amount of unclassified information, is available to DoD and contractor developers of models. Yet development of models of warfare that reflect Soviet perspectives is being carried out only for isolated projects. No development of models of nuclear warfare from an allied perspective is now underway.† Soviet and (to a limited extent) allied perspectives are treated qualitatively in selected parts of net assessments and some theater nuclear force modernization analyses, but the results of these qualitative analyses are rarely integrated with outputs from quantitative models.

As a notable example, most US models of nuclear warfare treat Soviet operations by using Western concepts, when in fact Soviet strategy for nuclear war is in many respects markedly different from that of the West. Some of the more important aspects of Soviet nuclear strategy

†*For example, models (except for some political-military games) generally assume that NATO functions as a single decision-making entity, not as sixteen sovereign nations in a loose alliance.*

that are reflected poorly or not at all in US models are: nuclear preemption;† surprise and deception operations; counter-C^3I operations; targeting of conventional and unconventional warfare forces against US and NATO nuclear forces as a matter of high priority; use of nuclear forces to support general-purpose military operations; treatment of conventional, chemical, theater nuclear, and intercontinental force operations as a continuum, not as distinct phases of warfare; and more extensive preparations than the United States or NATO for operations on an integrated battlefield.

Soviet concepts of operations are beginning to be incorporated in selected models (e.g., use of Warsaw Pact unconventional warfare forces in some models of theater nuclear force prelaunch survivability; models of Soviet civil defense efforts; modeling of Warsaw Pact echeloned ground force operation in some campaign models and tactical board games). For the most part, however, Soviet concepts of nuclear warfare are treated realistically only outside of quantitative models.

Criteria for Models of Nuclear Warfare. The foregoing review of recurring problems with models of nuclear warfare suggests a set of criteria by which model developers, analysts, and the users of analysis should assess candidate models and set objectives for improvement of nuclear warfare models. The purposes served by models of nuclear warfare should be to help:

- Expose weaknesses in force postures or weapon systems that may require attention.

---

†Preemption has a specific technical meaning in Soviet strategy. The Soviets apparently believe they will obtain advance indications of US or NATO preparations to employ nuclear weapons. Their doctrine calls for them to act on these indications and seek to preempt US or NATO use by striking first with Soviet nuclear attacks. See, for example, Benjamin S. Lambeth, "Selective Nuclear Operations and Soviet Strategy," in Beyond Nuclear Deterrence: New Aims, New Arms, Johan J. Holst and Uwe Nerlich, eds., Crane & Russak, New York, 1977, pp. 79-104 (esp. pp. 91-93).

- Compare alternative force postures, weapon systems, and targeting approaches in a wide variety of nuclear warfare scenarios to identify the relative strengths and weaknesses of each alternative.

- Achieve confidence that a force is sufficiently robust to provide adequate capabilities in a wide range of circumstances, especially in stressing situations.

- Synthesize new alternatives for examination.

- Provide systematic focus to debates about nuclear warfare forces and C^3I, and to bring discipline to these debates by representing important physical and operational realities quantitatively where possible.

These purposes imply that good models of nuclear warfare should strike a balance among (1) fidelity to reality; (2) provision of a wide range of nuclear warfare tests for forces, targeting doctrines, C^3I systems, and other support; and (3) enhancement of open and constructive debate as part of the analysis process.

1. Models of nuclear warfare should have adequate fidelity to reality, as required to support specific analyses. This means that models should:

   - Have good predictive capability in those many detailed areas where good data are available and the physical and operational relationships are well-understood.
   - Reflect important uncertainties and facilitate routine consideration of uncertainties when the models are applied.
   - Provide for routine consideration of two-sided use of nuclear weapons, integrated conventional-chemical-nuclear effects, important operational realities, and campaign outcomes vice summaries of casualties and damage

2. Models of nuclear warfare should make it easy to evaluate forces, doctrine, systems, and support in a wide-ranging set of circumstances. This means that models should:

   - Have broad scope to include a full range of scenarios and conditions deemed appropriate for a specific analysis; to embed nuclear forces routinely in a broader political-military context; and to carry out analyses from Soviet and allied perspectives, as well as those of the United States. There generally will not be a "best" single model for a class of nuclear warfare problems; complex issues need illumination from many different aspects.

- Be easily adaptable to analysis of new nuclear warfare issues, to the interests of different Government organizations, and to exploration of new insights emerging from ongoing analyses.

3. Models of nuclear warfare should facilitate and enhance open and explicit review, debate, and dialogue as an essential (as well as inevitable) part of the analysis process. Therefore:

- Models should be transparent to both analysts and the users of analysis. That is to say, models should make it easy for analysts to make it clear to users what are the major factors (assumptions, data elements, algorithms, scenarios, etc.) that determined the results of a particular analysis or output from a model. Model transparency will also facilitate peer review of models and analyses. The requirement for transparency should not, of course, be an excuse for using a simple model that does not capture enough of reality to be helpful.
- The organizational structures within which models of nuclear warfare are developed and applied should facilitate explicit review and debate about these models and their outputs as an important means of model validation.
- Models of nuclear warfare should be reproducible in the broad sense discussed below.

These criteria indicate that models should be clearly subordinate to the analysts of nuclear warfare and the logical framework they develop for approaching a specific issue. This is probably true for any field of inquiry, but especially so for issues about nuclear warfare, where the issues are highly complex, there is a paucity of historical experience with many aspects of nuclear warfare, and the consequences of choices are potentially enormous. The above criteria also indicate that a wide range of models should be available to allow the analyst to select appropriate model approaches and level of detail for application at discrete points in the logical framework for a specific analysis. This family of models should encompass gaming techniques, simple mathematical models, detailed computer algorithms, and simulations; models should be capable of modular application, so the analyst can easily tailor the model support to the specific issue at hand. Nuclear warfare issues are far too varied and complex to be treated with a single, all-purpose model.

"Reproducibility" as a criterion for models means the ability to obtain the same results from repeated applications of the model using the same inputs. Models of nuclear warfare should be reproducible in a broad sense that does not discourage analysts from using non-mathematical or partially mathematical models such as scenario narratives or gaming, when appropriate. If part of a model is judgmental, then applying the same judgments should lead to the same results. If a model uses human players, then the same human decision inputs should lead to the same model outputs. Reproducibility is important to help ensure there are no hidden factors operating in the model influencing results in ways that have not been made explicit. While reproducibility of a model does not necessarily result in greater fidelity to reality or better support for decisions, it does aid in providing transparent analyses and lending credibility to the model. But reproducibility should not be demanded in narrow ways that force the analyst toward exclusive use of mathematical models or computer simulations, with loss of essential, but difficult to quantify, dimensions of nuclear warfare.

CURRENT STATE OF NUCLEAR WARFARE MODELING

What are the current trends in modeling of nuclear warfare? What substantive areas are modeled adequately, where are development efforts currently focused, and what important areas are receiving little or no attention in model development?

Modeling of many basic elements associated with strategic and theater nuclear exchanges is, in a relative sense, more adequately done than for other areas of nuclear warfare analyses and, in an absolute sense, is probably sufficient for many analyses. These elements include the prelaunch survivability of fixed-base nuclear systems (e.g., ICBM silos, air bases), the allocation of nuclear weapons in near-optimal and operationally realistic ways (when the payoff function and allocation constraints are well-understood), and the calculation of damage to fixed and mobile targets. Important aspects of collateral damage (such as prompt effects and fallout) are reasonably well-modeled, as is the connectivity of communications systems in various attack environments.

Models of the penetration of ballistic missile and air defenses are adequate for many problems, although improvements are needed in critical areas of air defense penetration models, such as the effects of electronic warfare and the effectiveness of air defenses against vehicles with low radar cross sections flying at very low altitudes. While improvements continue to be made in models of these classical elements of exchange calculations, the physical and operational relationships are reasonably well-understood; data bases of reasonable validity are available; uncertainties are reasonably well-quantified; and the models permit sensitivity analyses and uncertainty assessments to be carried out. Modeling of these basic exchange calculation elements is an essential ingredient of nuclear warfare analyses, but addresses only a small part of the problem.

As discussed above, it is imperative to model -- and model adequately -- a far broader set of military operational phenomena in order to perform analyses relevant to current nuclear warfare issues. Model development and improvement efforts are underway in a number of these areas, with varying degrees of progress. Many of these efforts are embryonic, but they are beginning to receive attention from model developers. Important examples of these model improvement areas include the prelaunch survivability of mobile nuclear delivery systems (e.g., MX in its multiple protective shelter basing mode, Pershing, GLCM); the endurance and reconstitution of forces and C^3I after some damage has been inflicted; operations on an integrated battlefield; secondary collateral effects from nuclear attacks (e.g., the effects of urban fires); the effectiveness of a variety of civil preparedness measures; and economic recovery of a nation after a large-scale nuclear war. Some attention is being given to a wider range of general war scenarios, and selected efforts are being made to model and game prolonged nuclear conflict. Models of theater campaigns are beginning to reflect Soviet concepts of conventional, chemical, and nuclear operations, and a few organizations are attempting to develop Soviet-style models of nuclear warfare.

Yet other areas crucial for nuclear warfare analyses are currently modeled poorly or not at all, and model development efforts have either not been attempted or have made slow progress. Some important examples include social cohesion in a nuclear war, the effectiveness of the military command process in a nuclear war, the psychological effects on combat troops of nuclear or chemical attacks, and the termination of a nuclear war. Another crucial area is the translation of the effects of nuclear attacks on military forces (e.g., casualties, damage, disruption, delays) into impacts on the outcome of the military campaign. This is also a problem in modeling conventional and chemical attacks on military forces. Some progress is being made on this problem, but much remains to be done.

Finally, in the area of model flexibility, there has been little progress in making quantitative models -- especially computer models -- easily adaptable to a wide range of analytic scenarios. Other tools (such as gaming, historical or logical narratives, and human judgment) are more readily adaptable to a wide range of scenarios, but there has been little progress on integrating the use of these diverse tools with the use of quantitative models. Integration of qualitative and quantitative models for nuclear warfare analysis is feasible today and is done infrequently for some research projects, but this integrated modeling capability is not routinely available to DoD staffs.

ASSESSMENT

To date, models of nuclear warfare have provided analysts with important technical support on a large number of detailed issues, but have not been nearly as helpful in treating the larger policy, employment planning, and force acquisition issues confronting national security decision makers and staffs. In some cases, models of nuclear warfare have been dysfunctional.

Models of nuclear warfare have been of value in defining nuclear warfare issues and identifying force posture and C^3I deficiencies. They have imposed some measure of discipline on nuclear warfare debates by keeping these debates rooted in some physical and operational realities. In many specific technical areas, models have enhanced analysis of and decisions concerning specific nuclear force acquisition issues.

In the 1960s and 1970s, however, there was an overfocus on computer models of nuclear warfare (especially models of nuclear exchanges) to the exclusion of other forms of models. The results were an excessively narrow range of analysis scenarios; neglect of many important operational realities; an emphasis on nuclear firepower and neglect of maneuver, tactics, command and control, and the time dimension of nuclear warfare; and the analysis of nuclear forces in isolation from other military forces and from the political-military context of general-purpose conflict. This overfocus on computer models of nuclear exchanges responded to the demands of the users of analysis; it was not a state-of-the-art limitation. But it was aided and abetted by model developers.

Computer models that came into vogue in the 1960s inhibited strategic thought by focusing DoD staffs on what they could calculate with these models -- stylized nuclear exchanges rather than the outcomes of military campaigns. Strategy began to move beyond nuclear exchanges to nuclear warfare in the 1970s, but the capabilities of DoD staffs and their supporting contractors for nuclear warfare analyses have lagged behind the movement of strategy. There has continued to be an overfocus on computer models and insufficient use of other modeling techniques, especially in the area of strategic force analyses. Because of the inherent limitations of complex computer models, adaptation of models to the nuclear warfare issues of the 1970s and 1980s has been slow.

Improvements are needed in many dimensions of nuclear warfare models. What should be the priorities for improvements? This is a complex question, but there are two general areas that should receive priority attention in the process of making specific improvements. One is to make a concerted effort to provide DoD staffs with the means and training to utilize and integrate such diverse tools as gaming, narrative analysis, simple mathematical models, computer simulations, and human judgment in analyses of nuclear warfare issues, and to do so on a routine basis. This will permit more immediate progress through fuller use of tools and techniques now available. The second priority area for improvement is model transparency, to make models and their applications more credible and useful to DoD decision makers. Perhaps the establishment of a user-oriented DoD Nuclear Warfare Analysis Center for support in applying these diverse tools would facilitate accomplishing these goals.

We conclude, then, with some guidelines for nuclear warfare analysts and the users of their analyses:

- Devote sufficient project resources at the outset to careful development of a conceptual framework for a specific analysis, including explicit identification and review of key assumptions. This is in contrast to the frequent practice of merely selecting a model from off the shelf.

- Many specific technical and operational phenomena are adequately treated with computer models. The analyst should ensure that the computer model captures the essential physical and operational relationships and that the input data are reliable. In these cases, discrete steps in the analysis can be treated with computer models.

- Analysts should, as a matter of routine, utilize other tools as well as computer models, using the human intellect and the organizational processes of peer review and structured debate to integrate the outputs of diverse tools into a coherent analysis. Candidate tools, in addition to computer models, include logic, simple mathematical models, history, scenario narratives, the various forms of gaming, net assessments of trends and asymmetries in opposing force postures, and human judgment.

- Insist upon transparency of all the modeling tools used in the analysis and include as a routine part of the analysis an explanation of what factors, data, assumptions, etc., were primarily responsible for the specific output of the analysis and how they influenced that output.

REFERENCES

1. *John A. Battilega and Judith K. Grange, The Military Applications of Modeling, Science Applications, Inc., Denver, 1978.*

2. *Garry D. Brewer and Martin Shubik, The War Game: A Critique of Military Problem Solving, Harvard, 1979.*

3. *Morlie H. Graubard and Carl H. Builder, "New Methods for Strategic Analysis: Automating the Wargame," Policy Sciences, vol. 15, no. 1, pp. 71-84, November 1982.*

4. *Francis P. Hoeber, Military Applications of Modeling: Selected Case Studies, Gordon and Breach, New York, 1981.*

5. *A.W. Marshall, "A Program to Improve Analytic Methods Related to Strategic Forces," Policy Sciences, vol. 15, no. 1, pp. 47-50, November 1982.*

6. *J.J. Martin and Douglas C. Olin, "Improved Methods for Strategic Analysis," Policy Sciences, vol. 15, no. 1, pp. 51-69, November 1982.*

7. *Martin Shubik, "War Gaming: For Whom and What?" Policy Sciences, vol. 15, no. 1, pp. 85-86, November 1982.*

DR. J.J. MARTIN

Dr. J.J. Martin is the manager of the National Security Studies and Systems Group with Science Applications, Inc. (SAI). He has directed numerous studies and written extensively on the subjects of nuclear strategy, targeting, force posture, and arms control; he worked in the Office of the Secretary of Defense during 1968-76, with responsibilities in these areas. Dr. Martin served as an officer in the US Navy during 1957-77, and earned a Ph.D. in Operations Research from the Massachusetts Institute of Technology. He has been active in MORS, where he was on the Board of Directors and served as Secretary-Treasurer. He recently directed a large SAI program for the Department of Defense on improved methods for strategic analysis.

ABOUT CHAPTER 3

More thoroughly than any other author, John Friel expresses the uniformed analyst's view of model building. No reader can read Colonel Friel's chapter without having a deeper appreciation of how important a knowledge of combat operations is to the construction of sound battle models at the micro (one-versus-one), medial (few-versus-few), and macro (force-versus-force) levels. As they say, "The devil is in the details." Friel draws on his vast experience with Air Force simulations to illustrate the pitfalls that face the unwary. Yet Friel believes with the rest of us that air combat is inherently less complicated than land or sea combat to translate into battle models! You will find that Friel concentrates on standing simulation models because they are what he knows and understands best. As you will see, they suffice to make his points.

Air Battle Models

by

John Friel

INTRODUCTION

The objective of this chapter is to make a potential user of analysis aware of the state-of-the-art of Air Force battle modeling. Examples will be drawn from theater force models, strategic bomber penetration models, and strategic missile exchange models. This will be done not by using specific model names, but rather by describing model characteristics. At any given point in time, there may be as many as four or five standing simulations available to address a given issue in the Air Force and contractor community. Although the models have similarities, they often have significant differences. The client needs to be aware of the characteristics that a model should have in order to produce the fidelity required from the analysis. Statements will be made that a particular issue cannot be adequately analyzed using a given-level model. The statements are not contrived for the sake of discussion, but rather examples drawn from the author's experiences with analyses that have been performed using models with less than appropriate characteristics.

ONE-VERSUS-ONE AIR COMBAT MODELS

The most detailed physical representations of the weapon systems in air combat can be incorporated at the one-versus-one level. At this level the physics of the aircraft, SAM, radars, antiaircraft guns, and air-to-air and air-to-ground missiles can be explicitly treated, although not all one-versus-one models do so.

In an aircraft, effects such as lift, thrust, drag, angle of attack, airspeed, acceleration, and how they all change as the aircraft maneuvers can be explicitly captured. Often the change in location of the vehicle center of gravity as fuel is burned or ammunition consumed is treated. Detection processes can be modeled as a function of onboard avionics, cockpit masking, size of the opponents, radar cross section, and infrared (IR) signature. Often a number of different radar cross sections and IR signatures are modeled as a function of the spatial relationship between the two adversaries.

This type model finds its maximum utility addressing questions of engineering trade-offs. For example, when contemplating the addition of leading edge slats to an air-to-air fighter aircraft, one has to consider the disadvantage of the extra weight and drag (which affects the aircraft through its entire flight envelope) versus the advantage of improved low-speed instantaneous turn rate. A well-constructed one-versus-one model can illuminate this issue. Another example of engineering trade-offs that can be examined at the one-versus-one level is a question of the benefit associated with carrying extra fuel or ordnance on an aircraft. Here it's a question of the effect of the extra weight and drag versus the benefit of the extra staying power or firepower.

The one-versus-one model is also a useful place to investigate the tactics and rules of engagement of one-versus-one combat which are inputs to (or constraints on) m-versus-m combat. For example, let's assume that a type "A" aircraft is going to engage a type "B" aircraft. A one-versus-one model can give insight into the various tactical maneuvers that each aircraft should attempt to use on the other. Furthermore, in the case of aircraft engaging a ground target, the tactics necessary to successfully attack a particular ground target can be evolved if you know the nature of the defensive environment that the aircraft faces. It's also possible at this level of modeling to determine the specific maneuvers necessary to defeat a particular type of surface-to-air missile.

Care must be taken in one-versus-one modeling, however, when we are comparing a single combat element that is intended to be employed alone with a combat element that is intended to function as part of a system. For example, a strategic bomber may be employed in combat as a single entity, whereas a given SAM that may engage it is part of a larger air defense system. The element intended for single employment, the bomber, typically has all the faculties available that the military planner felt were necessary to carry out the mission. On the other hand, since the element that is merely a part of the system was conceivably intended to contribute to some synergistic effect of the total system, its effectiveness is understated if it is considered to engage by itself.

A similar situation can be encountered when one-versus-one models are used in quantity-versus-quality comparisons. The "quality" aircraft has more onboard capability in general. The "quantity" aircraft may depend on the existence of other capabilities inherent in some larger system. In addition, the larger numbers and usually higher sortie rate of the "quantity" aircraft are not captured in one-versus-one models. In general, one-versus-one-based analysis will favor the "quality" candidate.

Another area that must be treated with considerable care when we are using one-versus-one models is the way in which the decision maker in the system is modeled. For example, consider the case of an air-to-air combat where decisions regarding maneuvers against an intelligent enemy are made by a simulated pilot. Ostensibly, the simulated enemy pilot must also be free to maneuver in such a way so as to adapt his behavior to the combat situation as it develops.

This caveat is often violated in models of surface-to-air missiles versus an aircraft. Typically, the aircraft is constrained to perform a series of preprogrammed maneuvers and is not capable of reacting to the maneuvers of the missile. On the other hand, the missile's behavior is modeled so that it can adapt its flight path to the behavior of the maneuvering target. This can lead to inappropriate conclusions. The reason for having a pilot in the system is to get adaptive behavior into the decision making -- as opposed to the programmed decision making that limits a surface-to-air missile, regardless of its sophistication.

As stated in the Overview, models can be used to compare one system with another, but difficulty can arise when one tries to compute an absolute measure of effectiveness. This is the case when a one-versus-one model is used for computing inputs to higher-level few-versus-few or force-versus-force models. It is essential that the one-versus-one model be used in a way that is consistent with the details of the larger scenarios. One factor that must be given careful consideration is the duration of time for which the one-versus-one conflict is allowed to continue. Often the one-versus-one simulation is run for longer durations of combat than would be realistic when more aircraft are involved.

In addition to being able to explicitly capture the physics of the situation, another advantage of the one-versus-one model is that it requires less sophisticated intelligence information. The intelligence requirements at the one-versus-one level typically involve considerations of physical capabilities of the systems employed, with little concern for employment doctrine, tactics, training, mission, and the like. This simplicity in itself gives the one-versus-one-level model a firm basis from an intelligence standpoint.

Finally, one-versus-one is also the level of modeling which has the maximum amount of test data available for model verification. The ability to validate a model against actual system performance in the air is inherently much greater at this level than at any other level of modeling.

Although the one-versus-one models have many advantages because of the restrictive nature of one-versus-one combat, the types of questions that can be addressed are limited. If one is interested in the broader aspects of air battle, it is helpful to use the few-versus-few level of modeling.

FEW-VERSUS-FEW AIR COMBAT MODELS

As one would expect, few-versus-few modeling is more complicated. The first question that is pertinent is how many systems constitutes a few, and to what level of detail should each of the "few" be modeled. There are a number of factors influencing this problem. In the case of the air-to-air combat, the basic question may be the air-to-air employment concept of each adversary. For example, if the enemy tactical employment concept is to use a flight size of four aircraft, then the few-versus-few, air-to-air model should be capable of handling a flight of that many aircraft.

It is somewhat different for the air-to-ground model. Whether it be air-to-ground in strategic or theater-level conflict, the disposition of the ground forces basically determines the size of the few-versus-few engagement.

Consider a flight of four A-10 aircraft attempting to attack targets in a battalion deployed on the ground. The number of targets that the A-10s can engage in the ground force is influenced by the way in which the targets are dispersed over the terrain. The way in which the ground forces can engage the A-10s is determined by the range, capability, and location of their various surface-to-air defenses. Also, how many defenses can engage the attacking aircraft depends upon the tactics employed by attacking aircraft. If the attacking aircraft are able to ingress to the target area and egress at very low altitudes, or if they are able to maneuver sharply in the target area, then fewer of the surface-to-air defenses will be able to engage or even detect them.

Another way to consider the problem is to consider the edge effects. For example, suppose the analyst desires to use a model in which four A-10s engage a company of armored vehicles. If the deployment of the ground forces is such that surface defenses from other companies of armored vehicles are able to engage the A-10s when they are attacking, then the results can be biased considerably by the fact that the analyst has ignored the contribution of the defenses of the nearby companies.

It is at the few-versus-few level that one can first model the synergistic effects that evolve from a well-planned, well-structured defensive system. For example, at the one-versus-one level, the tactics that help protect attacking aircraft from an SA-6 surface-to-air missile could very well place the attacking aircraft in the heart of the engagement envelope of some other system in the defensive array of the ground forces. One gets a considerably different view of the effectiveness of the air defense when one considers it performing as an integrated system.

It is equally true that one gets a different impression of the effectiveness of air-to-ground aircraft when they are able to engage a system of defenses in a way that is mutually supporting. As the first aircraft rolls in to engage the target, the second aircraft may very well follow to engage any defenses that attempt to engage the first aircraft. The synergisms that exist in both attacking and defending forces become apparent in well-done few-versus-few models.

The intelligence information required to successfully use a few-versus-few model is substantially more extensive than for a one-versus-one model. For example, does an adversary commit his aircraft in flights of two or in flights of four, or in even larger groups? When enemy fighter bombers are attacked, do they jettison their ordnance and run, do they jettison their ordnance and fight, or do they retain their ordnance and attempt to continue to their target regardless of the attackers?

How are surface-to-air missile systems dispersed throughout the ground organization, and how do they engage attacking aircraft? Does the firing doctrine call for firing more than one SAM at the same aircraft? Does it allow for firing a number of different types of SAMs simultaneously at the same aircraft? How do the surface-to-air missile systems and the antiaircraft artillery systems interact in the engagement of attacking aircraft? In short, where the one-versus-one level merely required knowledge of engineering details of the systems, the few-versus-few level requires knowledge of engagement policy, training, employment doctrine, and command and control systems.

The question of command and control becomes relevant at the few-versus-few level. How are defenses internetted? For example, how many transporter erector launchers (TELs) are assigned to each acquisition radar? Does each TEL have its own tracking radar? How is the command, control, and communications (C^3) system connected? How are orders passed along the communications net? How does the C^3 system degrade as part of it is destroyed or if communications are jammed? What are the rules of engagement for the fighter? All of these questions become important when one is considering modeling at the few-versus-few level.

Once one introduces the notion of C^3, then countermeasures become important. For example, the use of electronic countermeasures to jam radars, the use of missiles to destroy acquisition or tracking radars, the use of communications jamming to force the SAMs and other defenses to behave autonomously all becomes a consideration at the few-versus-few level.

The decision process, although present at the one-versus-one level, is more complex at the few-versus-few level. In air-to-air combat, the simulated pilots must react to information from preflight instructions, from flight and element leaders, and from their own sensors. In addition, it is reasonable to expect the simulated pilots to process this information in accordance with how the fight was going. For example, if the fight started out with four aircraft on each side and one side is down to two aircraft because two have been destroyed, it would be reasonable to expect the actions of the two surviving pilots to be considerably different than they were when the fight was equal.

Another complexity that enters the modeling process at the few-versus-few level is the question of what is the appropriate measure of merit with which to summarize the results. In air-to-air combat, one could consider the number of friendly aircraft lost, the number of enemy aircraft lost, the fraction of friendly or enemy aircraft lost, the ratio of friendly to enemy aircraft lost, and so on.

The problem of measure of merit is also complex when one considers modeling few-versus-few, air-to-ground combat. Should the measure of merit be the number of tanks killed -- or the number of armored fighting vehicles, which includes not only tanks but also other vehicles? Should the number of surface-to-air defenses that are destroyed be part of the measure of merit?

Also, if some subset of the vehicles assigned to the organization is to be construed as the measure of merit, then one must consider the degree to which the attacking aircraft can ascertain the difference between a tank and an armored personnel carrier and a truck or a SAM at the range in which their weapons are employed. Care must always be exercised in considering the way in which the measure of merit influences the conclusions that are drawn.

The final aspect of the discussion of few-versus-few modeling is the question of how the model can be validated. Model validation for few-versus-few is considerably more difficult than it is for one-versus-one because of the shortage of data. However, in recent years the use of instrumented ranges and realistic scenarios in training and testing exercises by the Air Force has generated considerable data. Exercises like Red Flag provide information that help validate models and promote their use in gaining insight into the processes involved in few-versus-few combat.

FORCE-VERSUS-FORCE AIR COMBAT MODELS

Force-versus-force or (as they are sometimes called) campaign-level models represent the upper level of a hierarchically structured group of models. The force-versus-force model typically depends for its fidelity on the detail captured in the supporting one-versus-one and few-versus-few models. This detail can be aggregated into factors for use in force-versus-force models. As a result of this fidelity, there are a number of questions that can be examined using the force-versus-force model.

The first of these questions is the quantity/quality question. One-versus-one and few-versus-few models usually shed no light on the question of whether a few expensive aircraft are able to defeat a larger number of cheaper aircraft. The implication is that the "quantity" airplane has fewer subsystems and hence can be employed at a higher daily sortie rate. As a minimum, the force-versus-force-level model should have sufficient fidelity to capture the essence of the quantity/quality question.

Another area that can be examined using force-versus-force models is the appropriate force mix once you have decided upon the individual elements. For example, how many B-52 penetrating bombers, B-52 standoff cruise missile carriers, and B-1 penetrating bombers should be included in a strategic strike force? Typically, this is studied by defining the force mix to be considered and then comparing the effectiveness of that force mix with some other equal cost mix. In the theater force situation, similar analysis using force-versus-force models is done to determine the mix of air-to-air specialized, air-to-ground specialized, and general-purpose systems.

The reason that these questions can be examined by using force-versus-force models is that the question can be addressed by using results that have relative rather than absolute validity. Furthermore, it is well to remember that force-versus-force models cannot be legitimately used to gain insight into the question of how much is enough because of the absence of absolute validity not only at this level in the hierarchy but also at the one-versus-one and few-versus-few levels.

Although a number of important insights can be gained with force-versus-force models, there are some serious problems. The first is the amount of intelligence information that their use requires. Considerable detail concerning the scenario is required beyond the traditional questions of numbers of weapon systems and their characteristics, and initial location of military organizations. The one-versus-one level and the few-versus-few level are basically involved with questions of physical and tactical intelligence data. On the other hand, the force-versus-force level requires insight into the strategic philosophy and strategy of the opponent. These very esoteric considerations do not appear in the intelligence literature with any consensus in the community.

Another problem at this level regards the duration of the campaign to be modeled. The longer the period, the harder it is to retain fidelity of the model. As a war progresses, adaptive tactics evolve. Pilots on both sides attempt to deal with the realities of the combat they are experiencing by adaptive behavior. In short, they learn. This is a very difficult characteristic to model.

In addition, in a combat of longer duration, there is the question of changing the role of the forces. After the initial three of four days in a conventional war in Europe, the general-purpose fighters may be removed from the air-to-air role and employed in the air-to-ground role. The ability to perform this reassignment in real life is quite a bit more complex than is typically captured in the modeling.

Another important way in which duration of combat adds complexity is the process of transition from conventional to theater nuclear war. The decision to transition is a very difficult one to model since it involves some very high-level political decisions on the part of a coalition of forces.

Duration of combat also has a strong influence on the amount of detail required in the air base, supply, and maintenance systems of the force-versus-force model. If only a few hours of the war are to be modeled, then on the average not more than one sortie can be flown per aircraft, and details of the supply and maintenance systems are not required. However, since close air support sorties are flown in response to development in the ground battle rather than all at once, the aircraft shelters and revetments and the air base runway must be modeled. Furthermore these facilities must be at risk to enemy attacks, and the airfield's ability to generate sorties should be a function of how well these facilities survive the attack.

If the duration of combat is more than a few hours but less than about four or five days, then much more of the air base detail must be modeled. This includes, as a minimum, aircraft and maintenance shelters, runways, maintenance hangars, fuel, ammunition, and spare parts. Once again, the mode must allow these resources to be put at risk to enemy attacks. The ability of an airfield to generate sorties should be a function of how successful the attacks are. The upper limit of four or five days depends on how many spare parts and how much consumables are assumed to be on the air base at the time of the attack, together with the consumption rate of these resources.

If the campaign duration is more than four or five days, then the amount of logistics infrastructure modeling required increases dramatically. Ammo and fuel dumps, supply depots, maintenance and battle damage repair facilities, transportation nets, and even computer facilities (on which supply systems depend so heavily) will all have a major effect on sortie rate. The facilities must be at risk to enemy attacks, and the ability of the NATO system to generate combat-ready sorties will depend on success against these attacks. Very few, if any, existing force-versus-force air battle models provide this type of detail; hence theater analysis done for scenario periods in excess of about seven to ten days starts to lose validity rapidly.

These discussions concerning duration of combat have been couched in terms of theater force scenarios. Obviously they also hold for models involving strategic missile and bomber forces, although the times for which more infrastructure details are required are likely to be different.

Another important aspect of force-versus-force or campaign-level modeling is homogeneity of detail. In the modeling process, the model builder decides which elements of the process are important and includes them, while excluding elements that he construes to be unimportant. Once an element is included for consideration in the model, it's important that it be treated at the appropriate level of detail so as not to gloss over its true impact on the combat process. This question of homogeneous level of detail is very judgmental, as is the question of which characteristics of the combat process are important enough to be included in the model and which are not.

At the few-versus-few level, the idea was introduced that the decision concerning the measure of merit is a complicated one. This problem is exacerbated at the force-versus-force level. Consider the question of the effectiveness of strategic nuclear forces employed against the Soviet Union. There are a number of measures of merit available (such as hard target kill potential, equivalent megatons on target, or fractions of target base destroyed) that could be used as the measure in this evaluation. Each of these measures of merit has its own specific shortcomings and advantages. The problem is to understand the way in which the measure of merit biases the outcome obtained using the model.

The same problem exists when one is analyzing force structure at the theater level of conflict. One traditional measure of merit is the number of aircraft killed on each side or the number of armored fighting vehicles destroyed on each side. It may be considerably more important in conventional theater conflict to consider the degree to which the adversary's ability to deliver nuclear weapons has been reduced. This addresses his ability to escalate to the nuclear level, his ability to inflict damage at the far more intensive nuclear level, and indeed his propensity to cross the nuclear threshold in the first place.

AIR-GROUND BATTLE MODELS IN COMBINATION

Invariably in the evolution of a force-versus-force model or during continued use of such a model in theater-level analysis, the desire arises on the part of management to link or connect a force-versus-force-level air battle model with a theater-level ground battle model. Therefore, it is useful to discuss the ways in which ground combat differs from air combat and hence the way in which ground combat models can differ from air combat models. This is important to understand before large amounts of resources are committed to link up two big models or to use them jointly in an analysis.

To begin with, surface combat takes place in an environment that is much less homogeneous than is the air environment. Consider the ability to commit ground forces located at Point "A" to combat at Point "B"; this process could be heavily dependent on such factors as:

1. The transportation equipment assigned or available to the moving unit.

2. The intervening terrain.

3. The condition of the road net connecting Point "A" and Point "B".

4. The season of the year.

5. The climate and/or weather.

6. The distance between Point "A" and Point "B".

7. The amount and type of traffic recently using the road net.

Although new factors may have some influence on committing air forces located at Point "A" to combat at Point "B", air operations are much less dependent on this kind of environmental information. One way to characterize this difference is to say that the air as a medium is much less variable and hence far more homogeneous.

Another way ground battle models are more complex is in the sheer number of elements to be considered. An attack on the Soviet Union might involve several thousand aircraft and cruise missiles on our side and several thousand interceptors and surface-to-air missiles (SAMs) on the Soviet side. A conventional war in NATO might involve several thousand aircraft and SAMs on each side. By comparison, the number of crew-served weapons that would be employed on the ground in Europe could easily approach one hundred thousand on each side.

A third way in which ground combat is more complex than air combat is the degree of inter-engagement dependence. The status of a ground unit at the start of a given engagement is very dependent on the nature and outcome of its previous engagement. The number and type of surviving personnel, weapons, and supplies were determined in the previous engagement. The opportunity to reconstitute these resources depends on the outcome of the previous engagement. Although the posture of a combat unit is a function of the orders it receives from higher headquarters, the range of possible orders is largely determined by the outcome of the previous engagement. Finally, the terrain on which a unit will next be engaged is determined principally by the location, nature, and outcome of the unit's last engagement.

Although some factors from a previous engagement will influence the way in which a subsequent air engagement takes place, in general the air combat process is much less dependent on this kind of prior engagement information.

Another difference is the nature of the way in which weapons are employed. In the first place, ground forces tend to fight in organizational groups: a battalion engages a battalion, or a brigade engages a division. So it is appropriate to consider a battalion-level combat model or a brigade- or division-level combat model. This is, in general, not true in the case of air combat. An F-15 wing might be simultaneously engaged in three different parts of Europe in three different kinds of air superiority missions against elements from different enemy air organizations. An A-10 squadron might have one group of aircraft engaged over the North German Plain and another operating at the Fulda Gap. A B-52 wing might have aircraft which strike widely separated regions of the Soviet Union. In general it does not make sense in an air combat model to think in terms of wing-versus-wing or squadron-versus-squadron.

Another distinction between air battle models and land battle models is the firing process. Air battle models tend to have "point" firing processes rather than "area" firing processes. An aircraft uses its gun by pointing at or leading a target. This is true in air-to-air combat and also in air-to-ground attacks. Contrast this with the surface employment of a minefield, the field-of-fire responsibility of a machine gun emplacement, or the area fire of an artillery barrage. The increasing employment of guided munitions in both air-to-air and air-to-ground missions is consistent with the notion of "point" rather than "area" fire. Still further, in surface-to-air engagements, radar and IR-guided surface-to-air missiles (SAMs) and radar-directed gunfire are rapidly replacing antiaircraft barrage fire.

The final aspect of air combat models that is different from surface models is the way in which the major elements are attrited. An infantry company can exhibit substantially different effectiveness characteristics depending on which 10 percent of its officers and men are killed. A ship may sustain a fatal hit and yet fight on for a substantial period of time...recall the destroyer picket ships versus kamikaze planes around Okinawa in WWII. The death process for aircraft in air-to-air or air-to-ground roles is considerably more discrete. The state of an aircraft hit by a SAM, an air-to-air missile, or a burst of gunfire is much more likely to be two-valued (i.e., alive or dead) than would that of a ship hit by a missile or an infantry company hit by an artillery salvo.

Each of the characteristics discussed -- the homogeneous environment, the small number of fire and maneuver elements, the inter-engagement independence, the point-fire engagement process, and the two-valued kill mechanisms -- tends to make air combat easier to model to an equivalent level of detail. Great care must be exercised before attempting to connect up force-versus-force-level air and ground battle models.

Another approach which can cause difficulty is the use of a force-versus-force ground battle model to which some "consideration" of air power has been added by the ground battle modeler. For one thing, the analyst must be concerned with the sensitivity of the ground battle model to timing and mobility, not just to firepower. For example, a tactical fighter force can be employed in the morning in the Fulda Gap and in the afternoon in the North German Plain. Such an employment flexibility can only be captured in a combat model that correctly treats the maneuver of forces as a function of time.

The problem arising from the marriage of air and ground combat models at the theater level is most severe for the close air support (CAS) and battlefield interdiction (BI) missions. These missions represent an important linkage between the air and ground battles and require careful consideration by the modeler.

Timing problems can arise from the linkage of an air battle model and a ground battle model. The time constants that characterize ground force movement, engagement of ground forces, and reconstitution of ground forces tend to be longer than the time constants that describe these same processes for air forces. If a ground battle model is built to account for the slower ground processes, it may not be able to reflect adequately how during the course of one ground engagement the air forces may have engaged, reconstituted, and reengaged three times.

In one ground battle model that has been used for some time, the targets destroyed by CAS and BI sorties were removed from the ground order of battle once every 24 hours, at midnight. Targets destroyed by ground forces were removed at the time the destruction took place. This approach severely understates the responsiveness of theater air forces.

The approach was used to avoid the modeling problem of double killing of targets -- i.e., having a CAS sortie attack targets already killed by ground fire or ground forces attack targets already killed by CAS sorties (both of which occur to some degree in real combat). This highlights the tendency to model some of these complex command and control processes with convenient but overly simple modeling techniques.

A further complication along these lines concerns the question of allocation of CAS and BI sorties to various locations across the theater. The linkage between ground and air combat models usually requires the development of some set of rules for allocating CAS and BI sorties to various parts of the battle. In general, it is best to avoid allocation rules that are based on planning factors. It may be appropriate to plan to provide a certain number of CAS sorties for each engaged division, or for each kilometer of front line. However, to use these planning factors for allocation decisions seriously understates the ability of air forces to mass quickly in response to developments occurring in the ground battle model. It is much more useful to have the air resources allocated as a function of the progress of the ground battle.

INTERCONTINENTAL NUCLEAR WARFARE MODELS

It has been argued that the issue of force-versus-force-level air and ground combat models is a complex one that should be approached with great care if responsible analysis is going to result. The same kind of care must be employed when considering force-versus-force modeling of intercontinental ballistic missile forces. Many ICBM models that purport to represent force-on-force are, in reality, simplistic expected-value aggregations of a one-versus-one computation. For example, knowing the yield and circular probable error (CEP) of a reentry vehicle (RV) and the hardness of the target silo allows computation of a single-shot kill probability. Next, knowing the desired level of damage, it is possible to compute the number of RVs required per target, compounding kill probabilities on the presumption of independence. It is simple accounting to establish for a given-size attacking force how many of the target force will survive.

The above model is troublesome. It is not able to treat footprint constraints on the attacking RVs, fratricide of the attacking RVs, the delicate timing of the attack required, tactics such as launch under attack or launch on warning, and problems of the attacker's force allocation. All of these shortcomings of the model result in overstating the effectiveness of the force which strikes first. Yet much of the work published in the unclassified literature is done with this very simplistic type of "force-on-force" model.

A considerably more complex model used for ICBM computations is the type that does some sort of one- or two-sided, optimal or quasi-optimal allocations of the ICBMs over the target base. Typically, the models seek to maximize some measure of merit subject to a system of constraints and the results of the opponent's attack. As in the previous case, this class of models often does not consider footprint constraints, fratricide, attack timing, or tactics. In addition, many of the subtle nuances required by the United States Defense Guidance are not treated.

The most competent missile effectiveness models are capable of treating this long list of important facets of the environment of the employment of ICBMs. However, there are very few models of this quality; they require extensive data bases; and they are among the largest and longest-running models in existence. Great care must be taken when interpreting the results of missile force exchanges produced with the less sophisticated models.

VERY AGGREGATED AIR BATTLE MODELS

Necessarily the process of going from one-on-one to few-on-few to many-on-many involves an aggregation process (i.e., as one goes up the hierarchy, one sacrifices detail in order to consider broader aspects of the process). There is often a tendency to want to go one level above the force-on-force or many-on-many level to what can be considered very aggregated air battle models. Often this process is accompanied by the use of some optimization algorithm that varies the employment decisions in such a way as to optimize some measure of merit. This may be done in a static way or it may be done over a period of time, as in the dynamic programming approaches that are used at the very aggregated level of modeling. Often the entire theater war, including air and ground battle, is treated by the consideration of just three or four variables describing each side. This is simply too high a level of aggregation to give any new insight to the experienced military analysts.

Another example of overaggregation is the use of Lanchester equations to describe campaign-level combat. It is important at the outset to distinguish between the simple Lanchester equations which are used to model air combat and the much more complex forms of "Lanchester-like" equations that have been developed for use with some force-on-force ground combat models. It is the former class that are referred to in the following discussion.

The basic difficulty associated with the simpler forms of Lanchester models is that the assumptions necessary all too often do not reflect the reality of the combat situation. Those hypotheses that must be satisfied to use the linear law are generally not satisfied in air-to-air combat. Those hypotheses necessary to use the Lanchester square law are typically not met in air-to-air combat either, with the possible exception of beyond-visual-range engagements among aircraft that have radar missile capabilities.

The troublesome part of the use of Lanchester-like models in this very aggregated way is that, independent of the situation, the answer is very sharply influenced by the form of the models selected. In a quantity/quality consideration the linear law favors the quality system, whereas the square law favors the quantity system. The outcome then follows somewhat automatically, even though neither choice represents the relationship between the realities of air combat.

Perhaps 10 or 15 years ago, some insights could be gained by using these very aggregated models. It would seem that the air combat modeling process has progressed beyond the stage where any new insights are to be gained from modeling the attrition process in such an aggregated way. It is conceivable that processes that have received less attention -- such as command and control processes, reconnaissance processes, and decision-making processes -- may benefit from some higher-level aggregated study. This suggests that the appropriate role for these highly aggregated models is diagnostic (i.e., to guide the analyst and model builder toward the proper construction and employment of force-on-force-level models, rather than to assist in the decisions themselves).

REFERENCES

1. *Steven L. Amdor, "TAC-Warrior -- A Campaign-Level Air Battle Model," Proceedings of the 37th Military Operations Research Symposium, 1976.*

2. *William T. Morris, "On the Art of Modeling," Management Science, Volume 13, Number 12, August 1967.*

3. *Lawrence J. Low, Theater-Level Gaming and Analysis Workshop for Force Planning (Volume I, 1977; Volume II, 1981), Office of Naval Research.*

4. *James G. Taylor, Lanchester Models in Warfare, ORSA, 1983.*

COLONEL JOHN FRIEL, US AIR FORCE

At the time he undertook this chapter, Colonel Friel was Chief Analyst for the Assistant Chief of Staff, Studies and Analysis, Air Force Headquarters (AFSA). Since 1971 he has devoted nine years of duty to building computer simulations, compiling data bases, conducting studies, and communicating the results to military and civilian leaders. At AFSA he had overall responsibility for model architecture for Air Force studies and analyses, and acted as the quality control point on major briefings and analyses, as well as in the studies contracting process. As an Air Force captain, Friel taught graduate courses in operations analysis, statistics, and quantitative methods and models. He holds a Ph.D. in Operations Research from Ohio State University. From September 1981 to July 1983, Colonel Friel was Commander,

Air Force Geophysics Laboratory, Hanscom AFB, Massachusetts. He is now Assistant DCS for Plans and Programs, USAF Systems Command. Colonel Friel is a member of the MORS Board of Directors.

ABOUT CHAPTER 4

Wilbur Payne devotes considerable space to discussing Army organization for model development and application. He also discusses the hierarchical structure of ground battle models and reasons therefore. While the Army scheme can hardly be called typical, Payne's detail is useful to convey that any Service's modeling work is a very big elephant, and most critics will have a feel of it that will be as limited as that of one of the blind men in the fable.

Not to be overlooked are other important points. Army models are developed primarily as aids in the planning process, but the training process is so tightly interwoven that some models serve both purposes. The wide variety of ground models differ in detail, which is helpful for cross-checking and corroboration. Payne's comparison of ground and air battle models, complementing Friel's, helps in understanding the properties of both model families.

In Payne's way of thinking, ground battle models are conceptually not different, either in design or application, from models in engineering or many other scientific fields. Differences in detail are generally traceable to the scale and diversity of the warfare process being modeled. Differences also result because opportunities for observation of combat are (thankfully) rare and experiments to verify the principal predictions of models are expensive.

Ground Battle Models

by

Wilbur B. Payne

The models discussed in this section are part of a larger family of models. Figure 6 shows a classification of the models in this family. Any of these could be called battle models. Even components designed for Army systems generally consider vulnerability to hostile fire as a design criterion. This paper will deal in the main with Category 6 in Figure 6, which is most commonly called Combined Arms Combat Models in Army terminology. The differences in names are more than cosmetic. Even though the purposes of Army models might be described as to aid the design of the ground component of future forces, some (if not all) models used for that purpose must include the effects or activities of the air component of the same forces.

At present this family of models can be described as a loosely-linked hierarchy. The intent of the classification system was, indeed, to display that hierarchical nature. The output of any of the first eight classes of models is, in some direct and important sense, input to other models in the family. For example, one typical use of Unit Operations Models is the study of the resiliency of small units in the force. To what extent can they maintain and regain the ability to perform their missions after receiving enemy fire? Unit Operations Models could be viewed as accepting parametric inputs to describe levels of enemy fire. So far as the model is concerned, one parameter is as good as any other. Nothing intrinsic to the model discriminates between one battery volley of fire from a 122mm howitzer and a billion volleys. Since the range of possibilities number from zero to infinity, it is best if the input to the Unit Operations Model is bounded. The reasonable bounds come primarily from the Combined Arms Combat Models, since these produce insight

1. ENGINEERING MODELS FOR COMPONENT DESIGN
   (e.g., HYDRAULIC SYSTEM DESIGN)
2. SYSTEM ENGINEERING MODELS
   (e.g., MISSILE SYSTEMS DESIGN)
3. WEAPON SYSTEM PERFORMANCE MODELS
   (e.g., VULNERABILITY ANALYSIS)
4. UNIT OPERATIONS MODELS
   (e.g., ARTILLERY BATTERY OPERATIONS)
5. FUNCTIONAL AREA MODELS
   (e.g., DESIGN OF AIR DEFENSE FORCES)
6. COMBINED ARMS COMBAT MODELS
   (e.g., EVALUATION OF ALTERNATIVE COMBAT FORCES)
7. COMMAND GROUP TRAINING MODELS
   (e.g., BATTALION COMMAND GROUP TRAINING)
8. MANAGEMENT MODELS
   (e.g., FORCE STRUCTURE ANALYSIS)
9. REAL-TIME PLANNING MODELS
   (e.g., LINE-OF-SIGHT ANALYSIS)

Figure 6. Classification of Army Models

into the intensity of fire to which various units might be exposed.

All examples of the ninth class, Real-Time Planning Models, are submodels or modeling techniques originally developed as research models. This hierarchical family is described as loosely linked because the various models in the family have been developed and are maintained and operated by a large number of different agencies. Because they have been developed independently and at various times with emphasis on the uses intended by the developing group, the interfaces between them are not smooth or automatic. There is almost always a labor-intensive step of statistical inference to translate the output of one or more of the models into the input required by another.

Even though the emphasis in this section is on the Combined Arms Combat Models, there is no effort to list -- much less compare -- the various representative models of this class. There are at least 100 such models in fairly frequent use within the defense establishments of the Western world.

As in other chapters of the monograph, this chapter undertakes to describe some features of the processes to be modeled as these relate to the leading characteristics of such models and the general use of such models in planning and management. The purpose is to provide some basis for eventual judgment of the adequacy of particular models for particular problems and some insight into what constitutes proper and improper design and use of these tools.

USES OF GROUND BATTLE MODELS

Principally the models discussed in this section support the planning processes of the Army. Given this, it is clear that they must be viewed as predictive models. In broad general terms, the factors that they predict may be described as the capability in combat operations of future forces and the requirements that sustained operations of these forces will generate for external support. They deal, at most, with a fraction of the total forces: the deployable force. They are almost exclusively used in a mode that compares the future consequences of alternative decisions to be made in the present or near-future time frame. In principle, the time frame of the predictive capability of these models should be as far into the future as the consequences of near-term decisions will have specific, predictable consequences. In practice, the particular models under discussion here are most frequently used to predict events in the 7 to 10-year future. Even when limited to 7 to 10-year forecasts, the results of a particular planning cycle involving exercise of these models are frequently upset by unpredicted technological developments, unexpected decisions by potential opponents, and even by broad changes in the general economic and strategic situation. Because of this, the larger planning process within which the models are exercised is necessarily iterative and, hopefully, convergent.

GENERAL SPECIFICATIONS FOR GROUND BATTLE MODELS

Some of the requirements imposed on ground battle models derive directly from the prior discussion of the role they play in the planning process. Others derive from certain general features of the process they intend to model. Key features can be seen by extending (and paraphrasing) the overview given in Chapter 3 on air battle models, which contrasts (in scope and scale) the differences between air and land combat as a process to be predictively modeled. The discussion begun there is extended here.

a. In the chapter on air battle models, Friel points out that the physical environment in which ground operations occur is less homogeneous than that for air battles. It should also be noted that elements of a ground force are much less discrete inhomogeneities in their normal environment than aircraft are in theirs. The consequences of this are that ground battle models must, in some cases, represent not only the macro environmental effects (such as atmospheric effects, terrain, and vegetation) which determine whether or not two direct fire entities could engage each other. They must also capture some micro environmental effects, usually described by the terms cover and concealment, which control detection rates and determine the frequency with which they may take each other under fire and the outcome.

b. Friel points out that the number of entities in a ground battle that might interact in some significant way is much larger than in air battles. It should also be noted that the entities are much more diverse in kind. The air battle it describes at theater level as involving only several thousand entities whose interactions are significant might realistically involve not more than 20 technically different kinds of weapon systems. A ground battle of the same scope would have at least one order of magnitude (and probably two orders) greater number and diversity of interacting entities.

c. Friel also points out that it is necessary to map the entities in a ground battle in a coordinate system that describes their location in an organizational framework as well as in an Euclidean space-time coordinate system. It is, however, something of an oversimplification to justify this with the statement "ground forces fight as organizational groups." Closer to the truth would be a statement that battles are fought by *ad hoc* task forces composed of elements of several organizational units of the larger force.

Because of this, it is usually necessary to describe two-sided combat operations in five coordinate systems: a single space-time coordinate system and separate ones that locate entities in an administrative organization (the peacetime TOE) and in an *ad hoc* organization for combat.

This is important because the natural relaxation times in ground combat seem to be minutes or fractions thereof. The task forces draw support from resources such as artillery, over which the force commander does not exercise control or enjoy exclusive use. The response times for such support are quite sensitive to the organizational structure, in particular to the command relationships and communication networks. These must be incorporated in the models since the organizational framework can be, and often is, changed during battle.

d. Friel further points out that ground battles have a higher degree of inter-engagement dependence. It should be added that ground battles have a significantly different intra-engagement structure. A particular aspect of this is in the degree to which ground battle models must represent logistic processes. Except possibly in a few cases where air-to-air refueling might be a significant factor, it is only at the campaign (say, numbered Air Force or Corps) level of air battle models that one must necessarily have close coupling between logistic and combat activities. In a ground battle of typical duration, even at the company level, logistic activities can have significant effects. Rearming and the prompt return to action of lightly wounded soldiers or lightly damaged weapons are examples. It is also probably true that decisions by higher levels of command have more direct influence during the course of ground battles. The tactical activities of small units and even individual weapon systems are determined partly by local decisions and partly by higher-level command. While the same is true to some extent in air battles, it appears that higher-level command controls a larger fraction of the activities of elements of the ground force. It also seems likely that command decisions of higher levels have more immediate and decisive impact on the outcome of sea battles than ground battles.

To put these observations into a semi-quantitative perspective, land combat at the theater level may typically involve some 50 division forces on one side and 100 divisions on the other. The significant activities of these forces may proceed in an area of 4×10^5 square kilometers and endure for several months. Each of the division forces will typically contain 2×10^4 people and have legal custody for 10^6 things. In a logical sense, the simplest model one could imagine would be a purely descriptive one which represented the location of the components of this force. Clearly, the output of such a model would not be transparent even if its input could be generated at any instant.

There is no evidence in history nor in our experience in developing land combat models and applying them to military planning problems that such models must capture and represent every observable aspect of the processes to be modeled. It is, in fact, explicit in most definitions of the word model that they are abstractions, representing only those aspects of the process that are important and relevant. That is the heart of the matter. The process to be modeled is intrinsically complex. The models must be predictive. They must treat some, but not all, observable aspects of the process to be modeled. The abstractions and approximations involved must be derived without appeal to a generally accepted or empirically validated general theory.

TYPES OF LAND COMBAT MODELS AND THEIR CHARACTERISTICS

The above general characteristics of land combat have led to the development of a family of models that includes specifically a hierarchical set of Combined Arms Combat Models. It is clear from the Overview to this monograph and in the chapters dealing with models employed in other functional areas of defense management that this is the same response that has been made to similar problems everywhere they have risen. In point of fact, except for differences in terminology, there seems to be a very high order of similarity of overall model strategy in all fields of endeavor that employ predictive models, whether these are defense-related or not.

Figure 7 shows one representation of the members of this hierarchy. It can be seen, as indicated above, that one fundamental division of this hierarchy is organizational echelon. The choice of echelons is not entirely arbitrary. Until the end of World War I, ground forces were organized in an alternating sequence of tactical and administrative echelons. Both of these exercised tactical control over subordinate units. In addition, an administrative echelon was organized to provide both administrative and logistic support to subordinate units. Tactical echelons provide great flexibility since they can, up to limits established by the span of attention of their commander, control arbitrary numbers of subordinate administrative units. Administrative echelons are a bit less flexible since their capacity to support subordinate tactical units will usually become constraining long before their span of control would be. This neat and tidy interposition of echelons of different kinds was greatly modified in World War II and subsequent organizations. However, there remain advantages to employing administrative echelons as the framework for the hierarchy of models, if only because their organization for combat is substantially more stable during a battle. In present organizational concepts, battalion, division, and corps are all administrative echelons.

No aspect of priority among the members of the hierarchy is really intended either by the structure of Figure 7 or the directions of the arrows. The particular arrows on the figure were drawn by someone who probably felt the most frequent and important use of the hierarchy was from the bottom up, to evaluate a new weapon or materiel system concept or design. In that class of problem, we usually begin with a translation of technical design specifications into operational performance estimates and pursue the evaluation only as far up through the hierarchy as seems necessary. But it would be equally possible to use the hierarchy top-down. That is, a theater-level analysis might provide some insight into deficiencies that can be pursued downward to identify the

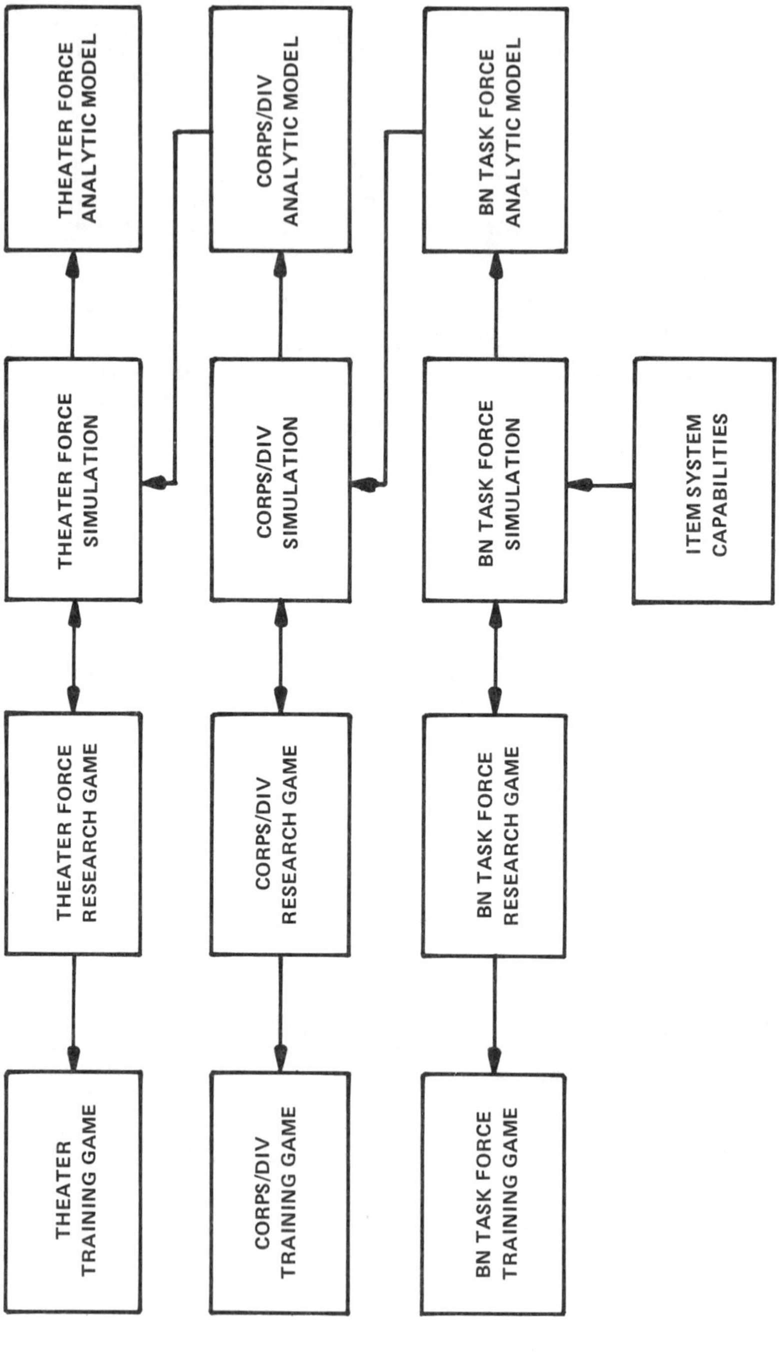

Figure 7. The Hierarchy of Combined Arms Combat Models

performance capability of materiel systems that might remedy them. Lateral and even diagonal movement is also possible. Ideas for improvement or the identification of problems that need attention can originate anywhere in the hierarchy, including from training games. Furthermore, the proper remedy for a problem identified somewhere in the hierarchy is as likely to be found through tactical or organizational changes as through new materiel development.

The vertical divisions in Figure 7 are a bit more difficult to describe. That division can be discussed in two different ways. First, it is possible to show that examples of such models already exist and, except for the set of simulations, have a long history. Admittedly it is not common to describe as "models" the command post exercises to train operating forces. But it would be hard to write a definition of "model" that excluded them. A second contrast among the columns of Figure 7 can be based on the fact that the purposes of, and the problems inherent in, design of the different types of models are different.

In common usage, two substantially different aspects are highlighted when an effort is made to distinguish "simulations" from other models. One of these is that most simulations attempt to represent explicitly the stochastic variability of various events in battle and make extensive use of the Monte Carlo algorithm to this end. A second is that simulations attempt to represent (through their input or their program) the impact of decisions which (in the process being modeled) are made by humans.

Training games may contain a man-for-man replication of the actual command structure of the unit or units being trained. Research games involve human players, but the players are usually playing several roles. Only a small fraction of the command structure is present. Simulations are, in this context, models in which human players do not intervene. Command decisions that would in the real force be made by humans are instead made by program logic that, among other things, specifies goals and objectives of the forces and represents a doctrinal interpretation of the principles of war and of common sense. Humans intervene through design of this program logic and through adjustments to it after exercise of the model.

In addition to being useful in some planning problems for which simulation is either inappropriate or suspect, war games also serve to check on the abstractions and approximations inherent in simulations.

Analytical models are generally distinguished from simulations in that they avoid intrinsic use of the Monte Carlo algorithm. It is *not* true of all of them that they fail to represent the stochastic variability of events in battle. Some, at least, of the "analytical" models can compute more than the first moment of the frequency function describing battle outcomes.

As indicated by the sets of arrows in Figure 7, analytic models are frequently employed as the lower-level combat models internal to higher-level models. To perform this role and provide a useful model of higher-echelon operations, it is very desirable that such models have very short running time. This has always required that such models be a higher order of abstraction and/or accept more approximate representations of important processes than is common in their cousins, the simulations.

Research games differ from training games in two important ways. First, since research games are usually used in a mode to compare alternatives (concepts, tactics, organization, and systems are the most frequent variables), it is usual to constrain the decision space available to the players. For example, it is hard to make sense in the comparison of two campaigns if in one of them a commander is misled by an enemy deception plan and in the other he is not. Model designers frequently take advantage of that. In most research games the consequences of all logically possible decisions are not explored. This makes possible a higher fidelity in representation of the aspects of battle which are under examination.

Training games, by contrast, should permit a wider range of decisions and represent their consequences with enough fidelity so that players are penalized for bad ones. A second desirable feature of training games is that some of them should force decisions at a pace characteristic of combat operations. Research games can, within limits, proceed at a much slower pace and compute the consequences of alternatives in greater detail.

THE LARGER FAMILY OF LAND COMBAT MODELS

Figure 8 shows the set of Combined Arms Combat Models in relation to a larger family of models. All of these could be called ground battle models since every one of them represents combat operations of deployable forces. The models in the left column (called "evaluation models") could, in principle, be any of the various types of models in Figure 7. As implied in their description as functional system models, the family of models on the right typically represent combat operations of the forces very abstractly (frequently as streams of events input to such models) and represent the activities of single functional components of the force in greater detail than is possible in Combined Arms Combat Models.

OBSERVATIONS

It is not unreasonable to ask whether so elaborate and extensive a family of models is really warranted by the problems to be solved. The existence of so large a set of models presents a variety of serious management problems. For example, they are resource-intensive. Further, there is no small group of people, planners, or managers who would claim to be familiar in technical depth with all of them or even any substantial fraction of them.

It should be remembered, however, that this family of models in concept and reality have evolved over a 40-year period of continuous activity. This activity has, in a very real sense, been decentralized. The fundamental reason for this is that readiness for future war during an era of rapid technological change has every appearance of being a bigger problem than can be taken at a single bite. In even a small war there is more than enough work to go around. In spite of the obvious virtues of the total systems approach and the dangers of suboptimization frequently treated in the literature of operations research, it seems necessary to divide this problem up in several parts so that many minds can work on it in parallel.

The larger organizational and management structure of the Army is to a considerable extent divided functionally in a manner that largely parallels the structure in the family of models. In the main, the design specifications for and detailed development of these models have been primarily at the initiative and under the control of these functional managers.

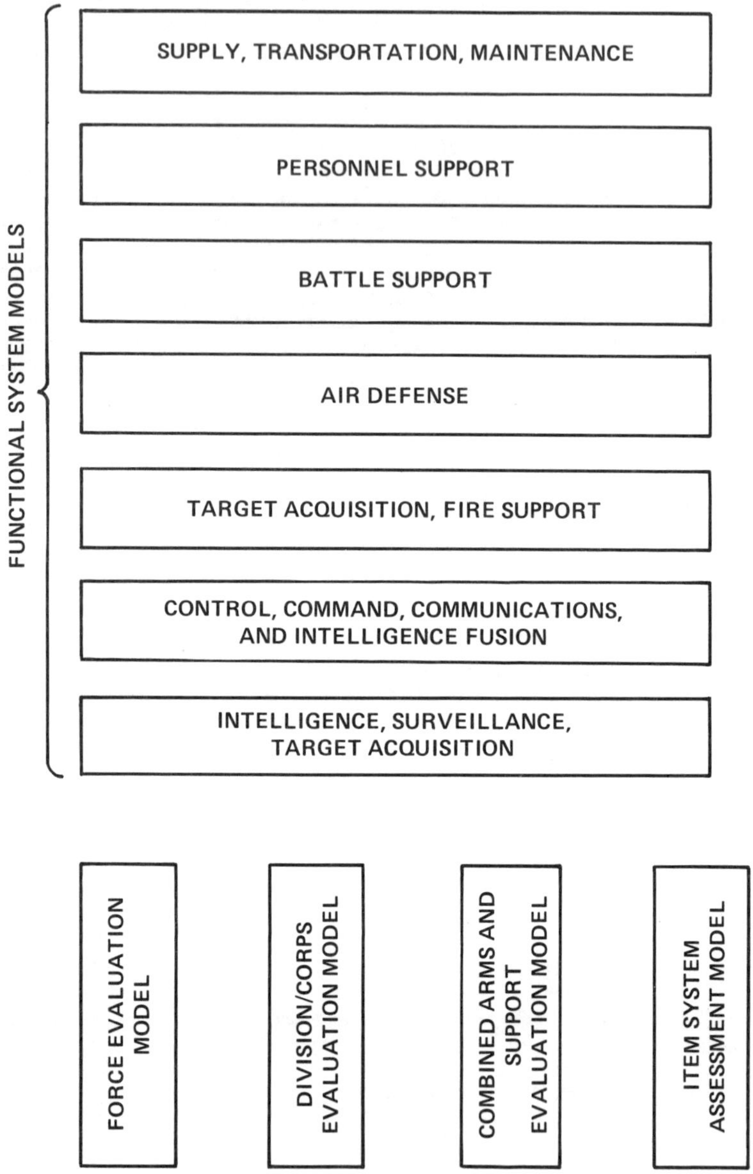

Figure 8. Family of Land Battle Models

Given, as discussed above, that each individual member of this family must involve both abstraction and approximate representations, there are almost certainly technical and management advantages to the multiplicity of models. Given, further, that they must be predictive models relevant to the consequences of near-term decisions, it should be clear that their predictions are not susceptible to direct empirical validation. Many but not all of their inputs could be said to have either an observational (historic) or an empirical basis. Some but not most of their forecasts could be (and occasionally are) adequately validated. At present and probably throughout the future, each element of the model hierarchy is itself redundant. There are, in fact, several quite different versions of each member of the hierarchy potentially available. This is sometimes viewed as unnecessary duplication of effort and unnecessarily costly. There is certainly some truth to that viewpoint. But from the viewpoint of questions of validity, this redundancy must be seen as a strength. The various models include different judgments about which factors can be safely ignored. They also contain different mathematical structures, differing degrees of approximation in their computations, and differing levels of detail in representing interactions. However weak it might be in an epistemological sense, the consistency or lack of consistency in output predictions is, in most cases, the basis for accepting or rejecting model findings and, in the case of lack of consistency, the best guide to learning more than we already know about combat. In this regard it should be noted that model findings may relate to decisions on the need for one or more experiments as well as to substantive program decisions.

TECHNICAL HISTORY OF MODELS

The intellectual history of war games has very deep roots. But the technical roots of the present family of models extend back only to the mid-1950s. As with nonmilitary models, the family of ground combat models derived from scientific and technological factors. The technological factors of importance in the 1950s were the increasing speed, size, and availability of the digital computer.

In the set of small-unit combat simulations, CARMONETTE and her myriad daughters, there were two elements of science. The first was a realization that the process to be modeled had some close kinship to the theory of time-dependent queues. Such problems can seldom be solved in closed form. However, from another branch of science the Monte Carlo procedure had evolved as an alternative to multiple numerical integration of stochastic processes. Thus, today's small-unit simulations could be described as the offspring of a *ménage à trois* whose members were a new insight into the process to be modeled, a new mathematical method, and the new technology of digital computers.

The analytical models of the operations of larger forces were all initially war games and evolved almost exclusively from the technology of computers. War games such as KRIEGSPIEL and its offspring had been in use for more than a century, both as a test of operational concepts and as training models. In the most elaborate of these an umpire staff of 100 or more people, using manual table lookup methods, implemented the game rules or, as it is sometimes called today, the battle calculus that translated decisions by the players into battle outcomes.

Transferring this burden to a computer made it possible to develop games that could be supported by a staff of 5 or 10 people. That, in turn, made the tools compatible with routine use for operations research purposes.

The latest evolution of this began in the mid-1970s and led to closed simulations of operations at division and higher levels. In these not only the battle calculus but also the tactical decisions were transferred to the computer.

NEAR-FUTURE IMPROVEMENTS

The present ground battle models have been described earlier as loosely linked. It has been known for many years that this linkage was important. The design of an inventory system involving the acquisition, storage, and distribution of supplies to support sustained combat operation requires, *inter alia*, information about the spatial and temporal distribution of customer demands.

The implementation of this desirable linkage has been informal, slow, and labor-intensive. Typically results from several applications of one or more division-level models would be gathered in written report form. The several applications and model would typically provide information on some but not all of the facts and factors needed as input to, say, an inventory theory model. The partial information contained in each report would naturally be compiled or scattered in form and format relevant to the purposes of that report. Some fairly primitive techniques of statistical inference would be applied to convert the output from existing reports into input to the next phase of the planning process.

Before one jumps to the obvious conclusion, he should remember that both design and execution of these models were decentralized. The various agencies responsible are widely separated both physically and organizationally. There have been occasions in which *ad hoc* task forces or special study groups have been assembled to address some particularly important or urgent planning problems. In these cases it has proven both possible and beneficial to attack a planning problem with a much tighter linkage among models. But this is somewhat clumsy, and the linkages remain labor-intensive.

Several factors make it appear feasible to produce a tighter linkage between existing models and to design families of models that can have quite tight linkages. The reduced cost of quite high-powered computers and the increasing availability of high-data-rate, secure communication nets is one such factor. A more subtle one is a radical shift in relative cost of hardware and software. Twenty years ago the cost of hardware was probably 80 percent of total cost in a large computing facility. Today that has probably reversed. In response, even where model development and execution remain largely decentralized, and even if hardware selection and procurement were similarly decentralized, there are very strong incentives toward hardware compatibility. The benefits of easy software exchange between several strong research groups now greatly outweigh any benefit from keeping up with the hardware state-of-the-art. And this is true even if viewed solely from the viewpoint of a single research agency.

This has led to an effort to build or modify models which, in their original design, recognize that the output of one member in the family is to be used as input to others. This is by no means properly viewed as a centrally managed effort to develop a single, super model. There is little reason to believe such a super model would be feasible with existing or forecast hardware capability, nor that it would be useful if feasible. Instead it is a centrally guided process through which agencies responsible for model development, improvement, maintenance, and execution accept some externally developed guidance and constraints. In the main, the constraints are in programming standards and formats and affect mainly the pre- and post-processors of models. The guidance is principally intended to insure that each member of the family produces a larger fraction of the input needed by the others and produces it in a form more easily used by the others.

The formal process through which this is being done is the Army Model Improvement Program. Initially it focused on the parallel design of three new Combined Arms Combat Models: CASTFOREM at battalion level, CORDIVEM at corps and division levels, and FORCEM at theater level. In parallel with these are efforts to develop better procedures for data base management. Since these three models are nearly finished, the present effort is expanding to develop the interfaces between these models and more detailed models of the logistics functional area.

FUTURE PROSPECTS

Of the three factors that permit improvement in models or modeling techniques, only the continued improvement in digital computers need be considered here. Wars are fortunately infrequent and, as is well-known, surrounded by fog. Mother Nature yields her secrets grudgingly. Deeper future insights into the processes being modeled or the evolution of dramatically more powerful mathematical procedures are predictable only if one puts no limit on the time to their occurrence.

Probably the technical area from which the third generation of ground combat models will come is the same as that from which the "fifth generation" of digital computers is evolving. That is progress toward solving the hardware and software problems inherent in the parallel use of hundreds of microprocessors. It is clear historically that the process to be modeled involves the parallel operation of many similar components. It is also clear that, at least over short intervals of time, the activities of fighting components widely separated in space are independent. If that is true, then it should be possible to restructure most of today's models for execution on parallel processors and, in doing so, to represent the actual system with higher fidelity.

DR. WILBUR B. PAYNE

Dr. Payne is the Director of the TRADOC Operations Research Activity headquartered at White Sands, New Mexico, and is the general manager of its three branches. He is, by education, a theoretical physicist with an undergraduate degree from Tulane University and graduate degrees from Louisiana State University. Former activities in military operations research include employment in the Operations Research Office of Johns Hopkins University, in the Systems Analysis Division of OASD(C), and in the Office of the Secretary of the Army as Deputy Under Secretary for Operations Research. Dr. Payne is also on the MORS Board of Directors.

ABOUT CHAPTER 5

"Sea Battle Models" by Robert Hallex is the last in our quartet of chapters on battle modeling. Hallex points out that for battle *planning* the Navy relies heavily on fleet exercises, but is now giving greater attention to war gaming. In contrast, analytical models and, to a lesser extent, computer simulations dominate in the weapon selection and force sizing decision processes. For these two purposes the Navy has maintained few standing models, usually tailoring a newly constructed *ad hoc* model suitable to address the issues at hand. In naval warfare, the trend in analytical and simulation modeling is toward simplicity, for reasons which Hallex discusses.

A particular strength of Hallex's discourse is an extensive, detailed description of the Navy's family of phenomenological, tactical engagement, single mission, and campaign models.

Sea Battle Models

by

Robert A. Hallex

INTRODUCTION

The purpose of this section is to describe the current state of development of sea battle models, their leading characteristics, and the general use made of these models. The discussion concentrates on two broad categories of models which, in both structure and application, frequently contain common elements:

1) Interactive war game models which test the skills and capabilities of two or more opposing forces engaged in relatively unstructured "freeplay" crisis or combat situations and,

2) Analytical models designed to evaluate the relative capabilities of alternative forces and weapon systems over a wide range of highly structured tactical situations.

As a generalization, models for naval battle *planning* include both fleet exercises and interactive war games, while analytical or simulation models have been employed more extensively for *weapon selection* and *force size/force level deliberations* in conjunction with the defense programming process. There is, however, an emerging trend toward integrating the results obtained from fleet exercises and war games into analytical models as well as using analytical models to plan and to evaluate tactical interactions as they occur in exercises or war games.

NAVAL EXERCISES

Interactive war games, ranging from relatively simple two-sided desktop games to extensive fleet exercises, are used for a variety of purposes including battle group training, developing/testing tactics, and naval battle planning. Naval exercises usually emphasize at-sea tactical training rather than investigative purposes, although there have been some extraordinarily well-conceived and executed fleet experiments in the sense described by Dr. Bryson in Chapter 11. For example:

- The UPTIDE Series in the Pacific in the 1960s which investigated carrier battle group effectiveness and survivability against submarines.

- The BIG and LITTLE DADDY Series in the 1960s which determined submarine-versus-submarine performance and developed tactics.

- The SHAREM Series in the 1970s for the same purpose in surface ship-versus-submarine engagements.

- The LANTCORTEX and SEACONEX Series in the early 1970s which evaluated the Sea Control Ship concept as well as the use of submarines in direct support of a task group.

However, most fleet exercises have been a compromise between personnel or tactical training on the one hand and data gathering, tactical development, or weapon systems concept investigation on the other. This compromise is due to the fact that fleet experiments are both extraordinarily time-consuming and expensive. Therefore, a great deal of attention must be paid to planning the experiments built into the exercise, and to the post-exercise analysis, in order that the lessons learned may be evaluated independently of exercise artificialities and the personalities involved. For example, the SEACONEX experiments demonstrated the effectiveness of the Sea Control Ship, but many knowledgeable observers thought that its success in those extended tests was more attributable to extraordinary leadership and the dedication of the participants than to any inherent merit in the concept.

In addition, fleet exercises must compromise between realistic freeplay and a desire to force interaction between participants. Although many exercise artificialities detract from realism, one of the most prominent is the forcing of encounters (engagement opportunities), and even the engagement interactions themselves, in order to accelerate the accumulation of tactical data. This in turn influences the encounter and engagement mode, timing, and engagement rate and range, all sensitive elements of modern naval combat. A second and related artificiality is the limitation on the physical size of the exercise area. The adequate representation of enemy raid size and weapon performance (air-to-surface missiles, submarine-to-surface missiles, or torpedoes) has also been a significant limitation of naval exercises.

A serious problem with many fleet exercises is that the imposition of data collection distracts the operators from full and complete performance of their duties, a most serious exercise artificiality. Automatically recorded data such as NTDS tapes, SINS position recorders, weapon officers printouts, etc., alleviate this problem. This automatic data collection, together with automatic reconstruction systems, has the potential to revolutionize both the conduct of exercises and the analytical objectives that the exercise can fulfill. Similarly, there are now several excellent test ranges where small-scale testing and experimentation with opposing forces in limited numbers may be conducted. Nonetheless, to extract valid and meaningful data from these or any other exercises, careful planning and significant additional resources (to gather data, "score" encounters for kill removal, etc.) are needed. Scheduling constraints, another resource problem, often causes exercise time compression, which further induces artificialities into the exercise results.

WAR GAMES

Recently, there has been a resurgent interest in war gaming for Navy battle planning. The Naval War College at Newport, R.I., of course, has been conducting naval war games for nearly a century. During the post-World War II period, however, with a few noteworthy exceptions, this gaming could be characterized as for training more than for investigation. With the establishment of a new Center for Naval Warfare Studies recently, it is expected that there will be a greater emphasis on naval warfare research, development, and battle planning.

Additionally, a renewed interest in war gaming is evident from a series of "seminar" games (a structured, time-phased discussion of a tactical problem), desktop games, and other war-gaming systems at Norfolk and San Diego and in the Fleet. Still moot is whether the current war-gaming emphasis will be upon training and execution or upon development and evaluation. In the mid-1970s, an Office of Tactical Development and Evaluation was established under the Chief of Naval Operations to improve fleet-wide coordination, but it would be overly sanguine to claim that it had put emphasis on modeling structure for fleet battle analysis.

MODELS USED FOR FORCE STRUCTURE PROGRAMMING

In contrast to the interactive models, a great deal of time and effort has been invested in the development and application of analytical and simulation models. These models have been developed almost exclusively as a decision aid in the weapon systems procurement and force planning processes. Since the advent of the current Defense Planning System in the early 1960s, virtually all Navy force structure decisions have been supported by extensive analyses of alternative sizes, mixes, and capabilities of weapon systems or combat forces. Sea battle models have been used extensively to provide the analytical framework in which weapon system capabilities or alternative courses of action and their potential consequences are evaluated. The principal use of these models has been to assist in the decision process rather than as a teaching or evaluation device for strategists or potential combat commanders. While occasional use is made of these models for battle planning or as a teaching aid, such use by the Navy has not been nearly as extensive as Army or Air Force use of similar land or air battle models.

REQUIREMENTS FOR AND USES OF ANALYTICAL SEA BATTLE MODELS

The program planning and review procedures currently used for Navy force-level planning is a highly structured process involving several Defense and Executive agencies as well as the Congress. Current and future force planning, formalized in the program planning process, includes a complete review of forces and capabilities annually.

Similarly, the weapon system acquisition process requires several formal reviews of each program during the research and development phases and during production. Both the program planning and the weapon system acquisition process require extensive operational analyses and cost-effectiveness studies to satisfy the Navy and higher authority that each new program, within constrained budget levels, meets one or several of the following criteria:

- Minimize military risk relative to threat.
- Maximize Navy capabilities to perform a wide spectrum of missions.
- Establish program priorities.
- Balance total mission requirements within available budget levels.

These operational analyses, based in large part on analytical sea battle models (along with suitable cost models), do not and are not expected to provide definitive numerical predictions of combat performance which may be directly translated into force planning decisions. Rather they are used as an analytical framework for the evaluation of relative capabilities of alternative force structures or characteristics. They do not, for example, provide definitive predictions of losses associated with maritime combat operations, since combat operations are heavily influenced by tactical posture, rules of engagement, and frequently by a substantial element of chance. Therefore, while sea battle models have proved to be indispensable for comparing alternatives, for identifying the factors which are dominant in determining outcomes, and for assessing the risks and uncertainties, at best they are merely analytical devices which assist in the evaluation of plans, programs, and procurement alternatives.

TYPES OF ANALYTICAL SEA BATTLE MODELS

Operational analysis studies based on sea battle models have proved to be both necessary and extremely useful in the force planning and review process. Unfortunately, the use of these analytical sea battle models for force planning can be quite expensive in both time and resources. There is a wide variation in resources required for different types of operational analyses and their supporting sea battle models. These models range from relatively simple tactical analyses (which use one-versus-one duel models) to exceedingly complex analyses involving major portions of Navy assets engaged worldwide in high-level contingency scenarios (such as those which include NATO and the Warsaw Pact) and which rely upon far more complex sea battle model methodology. The sea battle models associated with these analyses have two distinguishing characteristics:

- All employ threat-oriented models which evaluate the performance of opposing units engaged in tactical warfare.

- All include specific scenario parameters such as operational environments and rules of engagement.

The major problems in developing, selecting, and utilizing these analytical sea battle models are:

- To identify the issues which realistically can be addressed by models of varying complexity.

- To structure the analysis to address each issue at a reasonable expenditure of resources.

- To obtain data in a form suitable for model inputs.

The level and complexity of analysis must be consistent with the decision under consideration. Recent trends in sea battle modeling have been toward simplicity, transparency, and flexibility. For the purposes of discussion of sea battle models, the following (arbitrary) characterizations of analyses and types of analytical models have been chosen:

- *PHENOMENOLOGICAL MODELS* -- models of a single phenomenon (such as sound or radar propagation) or of simple observable or predictable effects (such as weapon/target interactions).

- *TACTICAL ENGAGEMENT MODELS* -- characterized by one-versus-one duels or many-versus-one engagements in a tactical environment.

- *SINGLE-MISSION MODELS* -- characterized by multiple systems engaging simultaneously or sequentially in a single mission or warfare area such as ASW, AAW, or STRIKE warfare.

- *CAMPAIGN MODELS* -- characterized by detailed scenarios and aggregations of several single-mission analyses which approximate the expected force interactions during a campaign.

Before describing specific examples of each of these types of sea battle models, a brief discussion of two factors -- environment and engagement rate -- which differentiate sea battle models from models of combat used for land and air warfare is appropriate.

ENVIRONMENTAL CONSIDERATIONS

In general, environment plays a different role in sea battle models than is typical of land or air battle models. For example, in models used for antisubmarine warfare, environmental considerations are absolutely dominant. Environmental factors unique to sea battle modeling are due to the opacity of the underwater environment, the extreme sensitivity of underwater propagation of sound to rapidly changing physical parameters of seawater (such as salinity, temperature and temperature gradient, water depth, marine life, and background noise), and the vast size of ocean areas in which both offensive and defensive forces may operate. The extensive size of operating areas in which a single submarine may be located, together with the range and destructiveness of modern naval weapons (both conventional and nuclear), and the key role played by surveillance are all important elements of the problem. A single nuclear submarine may operate over the entire North Atlantic Ocean, without resupply or support from any other unit, for an extended period of time (often several months). Even when naval units are deployed in mutually supporting groups, the number of ships and aircraft operating in mutual support tends to remain small relative to the area encompassed and to the number of units considered in land or air battle models. Even a major force element such as a carrier battle group would normally consist of fewer than 10 ships.

ENGAGEMENT RATES

In land and air battle models, the engagement rate between opposing forces is normally high relative to those used in sea battle models. Once engaged, land or air units (or parts of units) tend to remain engaged until an objective is attained (or lost). ASW naval campaigns, on the other hand, tend to proceed at a much slower pace because operating areas are vast and surveillance coverage is poor (especially underwater surveillance). Opposing forces tend to operate in small and highly mobile units.

Thus it is difficult for an area ASW commander to force a decisive engagement with a deployed submarine force that chooses to be evasive. Since engaging naval forces normally consist of 5-20 ships (rather than hundreds or thousands of units typical of air and land warfare), the outcome of naval engagements is likely to be quite decisive (i.e., engaging units are likely to be destroyed or heavily damaged or, alternatively, survive in fighting condition). Therefore, opposing naval forces could not tolerate continuous engagements typical of land warfare. For these reasons, naval battle models tend to concentrate on three separate engagement modes:

- Search and attrition operations over extensive operating areas. Operating areas may exceed a million square miles, while limited search areas covered by single units (such as a patrol aircraft or a submarine) may consist of several thousands of square miles of ocean. Detection of individual enemy units within an operating area may require weeks of search. Localization of an underwater target may consume several hours or several days before attack criteria are achieved. Mine warfare may be protracted and exhaustive.

- Continuous patrol over and defense of a more limited but often mobile operating area (usually that surrounding a battle group, convoy, or submarine/antisubmarine barrier). Major interactions between forces are likely to be few in number but quite intense and decisive when large forces engage.

- Prolonged operations (such as strikes ashore or amphibious assault) in relatively fixed operating areas which include fixed defensive perimeters against air, surface, or submarine attack. These long-term operations normally evolve rather rapidly into a predictable "steady-state" OPTEMPO. Long-term air strike campaigns closely resemble the air battle models described elsewhere but employ mobile bases and logistics. Shore bombardments and amphibious assaults use modeling techniques somewhat similar to the land battle models described elsewhere.

The combination of high mobility, complex surveillance, large operating areas, and small numbers of units included in sea battle models tends to cause concentration of attention on restricted ocean passages or "funnels" through which enemy forces must transit to or from operating areas. This effect is somewhat analogous to the defense of natural barriers such as critical mountain passes or rivers in land warfare, but with larger dimensions.

PHENOMENOLOGICAL MODELS

The simplest and best understood of the classes of models used for sea battle investigations are the phenomenological models. These models normally investigate only one or a few phenomena simultaneously. They usually are used to evaluate the effectiveness of sensors, weapons, or weapon systems relative to various classes of targets.

The complexity of detection models is usually directly related to the complexity of the phenomena under investigation. In the case of microwave radar performance, the environment is relatively homogeneous and target strengths well-known or easily estimated. The radar power equation provides a good estimate of signal strength, and a straightforward engineering analysis of the transmitter and signal processor yields a good estimate of detection performance. As the environment becomes more complex and multipath phenomena become a significant factor (as is the case with HF OTH-Backscatter radar), the models must be made more complex. This is particularly true when the environment is not only complex, but also highly variable, as in long-range underwater search. In this case, a good statistical model of the behavior of the environment is also necessary. Usually a highly reliable statistical model of the environment is not readily available, especially in the case of seawater sound propagation phenomena. The major difference between the phenomenological models used in sea battles and those used for land or air warfare is the pivotal role played by these complex propagation and detection models in *most* sea battle investigations. These complex propagation models are combined with tracking, maneuver, and homing models similar to those used for land and air combat.

Another use of phenomenological models is the evaluation of weapon performance against a wide variety of target classes. In these investigations, damage effects against land and air targets are evaluated in exactly the same manner as those described in the land and air combat models section; that is, damage is estimated as a function of aiming or homing error, blast effects, and target area and hardness. In sea combat models there is a much larger area of uncertainty involved in fire control solutions for underwater targets. The complex underwater propagation of sound phenomena (both for active and passive systems) must be fully accounted for during the target acquisition and homing phases, since this usually proves to be the dominant term in the weapon system performance equation. Although these models frequently include a great many performance parameters, much of the required data is available from test programs, well-known physical models, or simple kinematic maneuver calculations. The statistical data base available for use in models at this level of detail is far superior to that available for those sea battle models which use higher orders of data aggregation.

TACTICAL ENGAGEMENT MODELS

Tactical engagement models generally are used to evaluate the effects of one-versus-one duels (such as a single SSN transit of a barrier station patrolled by an opposing SSN). Similarly, a single onstation search period by an ASW patrol aircraft (in a single operating area) searching for enemy submarines would also be modeled as a one-versus-one situation. Fighter-versus-fighter or intercepter-versus-bomber engagements also fit this simple modeling technique. Data for all of these models are strongly supported by tests at sea.

These one-versus-one models frequently employ simulation techniques and usually contain a great deal of detail concerning the physical aspects of the environment as well as the performance characteristics of the opposing units. For underwater engagements, the characteristics of the acoustic environment are modeled in great detail, as are the characteristics of all sensors and processors which provide information to the commanders of opposing units. Details of platform performance (such as speed, turn radius, engagement envelopes, radiated noise characteristics, depth limitations, etc.) are considered. Performance estimates of weapons (such as torpedoes for submarines; SAMs, torpedoes, or cruise missiles for surface combatants; AAMs or ASMs for aircraft) are modeled in considerable detail. Firing doctrines, target aspect and vulnerability, maneuver tactics, and other relevant considerations must also be included in these models.

An extension of the one-on-one duel situation, which uses essentially identical modeling procedures, is the many-versus-one (or frequently many-versus-many) engagement model. Normally, all units in the surface screen protecting a battle group or convoy are modeled in much the same detail as a single ship. The characteristics of the entire screen are derived from evaluating the performance of individual units and their spatial relationships. The effectiveness of the entire screen (or several variations of the screen, including alternative component units or configurations) is then evaluated for a wide variety of target types and a representative spectrum of environments. Other combatant units mutually performing a single function (such as SAM screens defending against ASMs, fighter screens defending against aircraft or missiles, or minefields deployed against submarines or ships) are evaluated in a similar manner.

The fidelity of these models may be checked by comparing predicted values with the results of previous exercises or with special tests. Although there is usually only limited data available with which these benchmark checks can be made, there are normally a few data points which can be used to roughly calibrate the model predictions.

Models designed to evaluate the relative outcome of one-versus-one or many-versus-one engagements are widely used to measure the relative effectiveness of alternative sensor suites, alternative weapon suites, and the performance characteristics of competing platforms or weapons. They are also used extensively to provide input data to a higher level of aggregation mission or campaign models.

Since the performance characteristics of individual systems under investigation (as well as the spectrum of environmental conditions) are reasonably well-known, tactical engagement models are well-suited for large-scale mathematical evaluation and computer-aided simulation. Most large-scale simulations of naval combat are conducted with this type of sea battle model. Due to the wide variability of environmental conditions in which naval combat may occur, the variation in target signature and performance characteristics of threat systems, and the variations in naval weapon system characteristics under investigation, a great deal of parametric analysis is mandatory to determine the relative effectiveness of alternative tactics or weapon systems. The resulting outcomes of interactions between friendly and enemy weapon systems normally exhibit an equally wide variation in system performance. Therefore, it is absolutely essential that the analyst and client employing these results understand the sensitivity of predicted outcome to alternative environments and threat characteristics. The variability of performance of competing systems over an appropriate range of expected environments must be included in reported results. Some of the serious errors made in naval analysis have been due to investigations which have been confined to one or two environments or to an average or "nominal" environment which heavily favors a particular weapon system.

Examples of the applications of tactical engagement models include the selection of alternative ship characteristics such as ship size and speed, weapon suite performance characteristics, magazine size and load-out, size and composition of the aviation complement, characteristics of competing aircraft designs, competing advanced hull forms (such as SWATH, SES, air cushion, or displacement), and specific sensors such as sonobuoys and horizontal or vertical acoustic arrays.

SINGLE-MISSION MODELS

Single-mission sea battle models are exemplified by multiple systems (such as aircraft, area SAMs, point defense missiles, guns, active and passive EW systems, and decoys) engaging an incoming raid of ASMs either simultaneously or sequentially. The major difference between these models and the previously discussed many-versus-many tactical engagement models is the wide range and diversity of engaging systems employed in a single mission (such as air defense) and an extensive engagement envelope which may be several hundred miles in depth. Usually single-mission evaluation models consist of several layers of engagement envelopes. These layered engagement models are the so-called "meat grinder" models which successively decrease the probability of one or more enemy missiles, aircraft, or submarines surviving to engage a target as they transit each layer of defense. The performance of each individual layer of defense is normally estimated by use of the tactical engagement models previously described. However, several crucial effects not previously evaluated become important at this level of aggregation. These include:

- The effects of weapon system saturation due to raid size and coordination (timing).

- The effects of command and control on coordination between defensive systems.

- The effects of raid geometry over the battle space.

- The cumulative effects of attrition during the engagement.

For engagements which involve a large number of units that interact in a relatively short time span (i.e., minutes or hours rather than days), these four effects frequently dominate the results; therefore, they must be considered fully. Since most mission models include several layers of interaction by a wide variety of weapon systems, it has proven to be impractical to model each layer with the fidelity normally achievable in a detailed simulation while simultaneously taking proper account of the effects of firepower, saturation, coordination, and the kinematics of engagement. While there have been many costly and time-consuming attempts to include all these effects within a single large-scale simulation, none (to the author's knowledge) has ever proven to be completely satisfactory.

Modeling difficulties are due to the inability of the analyst to properly account for the overwhelming number of important tactical and environmental interactions, particularly when using Monte Carlo techniques. A more realistic modeling approach (and one which has consistently produced more tractable and analytically verifiable results) uses weapon system performance results obtained from lower-level models as input to more aggregated models which realistically account for the aforementioned tactical complexities. Care must be exercised to ensure that input data derived from lower-level models correspond with the situation under investigation in the aggregated model, since failure to do so almost invariably leads to erroneous conclusions.

Due to the wide range of Navy missions, the range of mission analysis models is also quite broad. Examples of multiple-mission models are the convoy or underway replenishment group protection models which involve both antisubmarine warfare and antiair warfare. These sea battle models include layered defenses consisting of surface combatants, submarines (direct support), land-based patrol aircraft, helicopters, and sometimes carrier-based aircraft. Battle-group support models also include combinations of these same forces, but in larger quantities (and usually opposed by a significantly stronger enemy force). Antisubmarine-barrier mission models include the combined effects of submarines, land-based patrol aircraft, and frequently CAPTOR mines. ASW area search or area sweep operations incorporate the effects of long-range underwater surveillance systems; coordinated area search or sweep operations by patrol aircraft, submarines, surface combatants, and carrier-based aircraft; and frequently include vectored intercepts in response to long-range detections.

In addition to providing baseline effectiveness estimates for use in the more aggregated campaign models, these mission models are used to compare the relative contribution of candidate systems in a mission area. These models have proved to be a powerful tool for trade-off analyses in cost and effectiveness between totally dissimilar systems (patrol aircraft, submarines, and surface combatants) employed in a single naval mission (such as area search or ASW barrier operations). Similarly, the performance of patrol aircraft, attack aircraft, patrol hydrofoil ships, surface combatants, and submarines can be compared in the antiship mission. Mini-campaigns involving fixed rules of engagement and a common geographic (and environmental) area are used when making these dissimilar system comparisons with mission analysis models. This form of sea battle model is frequently used for major weapon system selection studies. Two recent examples of analyses which employed mini-campaign mission models are the Force Mix (Convoy) Study and the Navy Fighter Studies, which estimated the relative cost-effectiveness of alternative ASW/AAW systems and carrier-based fighter designs, respectively.

Although the relative marginal contribution of dissimilar systems may be estimated with mission area models, it is important that the analyst be aware of several peculiarities of layered defense models. The order in which additional weapon systems are added to a mission area model often is more important than the relative effectiveness of the system. This phenomenon is usually due to the peculiarities of the mini-campaign scenario. As an example, the addition of a submarine screen to an ASW force may increase the expected number of kills by 50% if it were the first system added to the screen; the improvement in screen performance may be only 10% if other effective systems were included in the baseline force ahead of submarines. Thus it is entirely possible for submarines to achieve more kills at less cost than any other system (due to encountering targets first in the layered defense network) while simultaneously achieving the smallest marginal (incremental) increase in total mission performance if added to a preexisting baseline force. Although this modeling effect is appreciated by most analysts, studies occasionally have failed to acknowledge its significance. A second effect frequently disregarded by single-mission models occurs because most naval systems are designed to perform several missions (such as ASW/AAW/ASUW). They normally do not perform all missions equally well. Single-mission models are likely to overstate the relative military utility of single-purpose systems and understate that of multi-mission systems.

CAMPAIGN MODELS

The highest levels of aggregation of data in sea battle models are found in campaign models. These models usually synthesize results obtained from many different mission area and tactical models, each of which produces aggregated data which are used to evaluate separate portions of a major campaign. These campaigns and their associated models are characterized by detailed scenarios which include threat levels, tactics, overall strategy, missions and objectives of opposing forces, force dispositions, and rules of engagement. Tactical interactions are modeled by a detailed sequence of events developed by experienced naval planners. The sequence of events includes convoy and shipping schedules, intensity and duration of strike campaigns ashore, area search schedules, barrier operations, and enemy counterforce strikes which are postulated to occur during each phase of the campaign. A salient feature of campaign models is that they must be coordinated with the military planner since strategy, tactics, and employment of forces must be adaptive and altered throughout the campaign whenever the relative combat capability of opposing forces is altered, for example, by the effects of attrition.

The primary uses of these very aggregate models are the evaluation of potential outcomes of major campaigns and the identification of primary determinants of outcome. They are also used to identify major capability deficiencies relative to a particular strategy and to evaluate the likely impacts of major force structure changes. These scenarios and sea battle models also are quite useful for evaluating alternative force employment options. They are occasionally used in interactive war games as a teaching device.

As useful as they may be for the broad purposes previously described, campaign models (due to their complexity and great reliance on additional layers of models for input) are difficult and time-consuming to construct and to use. Due to the complexity of the subordinate models necessary to develop baseline performance estimates in each mission area and the wide variety of mission areas (each involving significantly different physical phenomena, system performance, and tactics), a great deal of skill and detailed knowledge of the various aspects of naval warfare is required. It is quite uncommon for professional naval analysts or naval officers to be expert in more than one or two of the large number of dissimilar mission areas which must be included in a worldwide naval campaign.

Due to the complexity of the models and the time required to conduct a coordinated campaign analysis, these highly aggregated models must be used with a great deal of discretion. Model results are known to be highly sensitive to tactics and timing. This is due, in large part, to the relatively small number of discrete units employed in each theater and the disasterous effects of losing primary units (one of two carriers out of action in a major theater may decrease total force capability by 50%). On the other hand, marginal contributions of small numbers of additional supporting units may be difficult to detect (one or two additional frigates in the Atlantic theater, for instance). Therefore, a campaign model is usually two coarse-grained to be useful in determining the marginal value of small force increments. But a campaign analysis is valuable in determining the number and mix of forces to be employed in the more detailed battle model. Due to the unique requirements for and uses of campaign models, their evolution and utility is greatly facilitated and strengthened by the continuity and stability of analytical teams. Complex campaign sea battle models have been used periodically by the Navy over the past 15 years, but little attention has been devoted to maintaining in-house analytical teams capable of exercising these complex models skillfully. A representative sample of the major analyses employing campaign sea battle models over the past 15 years is shown below:

STUDY	SCENARIO	CONTINGENCY YEAR
CYCLOPS I, II, & III	COMMAND OF THE SEA, ATLANTIC	1963-65
WAS-I	WAR AT SEA	1966
WAS-II	WAR AT SEA	1968
NARAC-G	NATO (WORLDWIDE)	1973
MEPEC-73	MIDEAST	1973
SEAMIX-I	NATO (WORLDWIDE)	1980
SEAMIX-II	MIDEAST/INDIAN OCEAN	1981
SEAMIX-III	SEA LANE PROTECTION	1981
TALLFORM-II	MEDITERRANEAN	1981
SEAWAR-85	NATO (WORLDWIDE)	1985
SEAPLAN-2000	WORLDWIDE	2000

SUMMARY

Studies employing analytical or simulation models of naval conflict are required and consulted at every major decision point in the Navy planning process, both for *force planning* and *weapon procurement* (from concept formulation through OT&E). The Navy has devoted less effort than the Army or Air Force to the use of analytical models or simulations for *battle planning*, but relies chiefly on fleet exercises and (increasingly) war gaming. Due to the frequent use of model results in the decision process, senior executives involved in the Navy management and planning process are relatively familiar with sea battle models and their results. Commanders at sea tend to be less familiar with these models and their results unless they previously have been a part of the planning and programming system.

Although the results of sea battle models are used extensively in planning, the Navy has not maintained a continuing modeling capability, but does its modeling principally on a decision-oriented *ad hoc* basis as a need arises. Little planning has been directed toward maintaining and improving models, and the Navy has no group of analysts dedicated to maintaining lower-level models that are designed to interface with complex campaign aggregations, either for battle planning or for a continuing review of Navy force programs.

Because of its *ad hoc* approach, the level and complexity of the Navy's sea battle models (fleet exercises excepted) tend toward simplicity: analytical in preference to simulation; simulation in preference to war gaming. The level of model aggregation is normally the lowest which will serve the purpose. For most Navy analyses, the proper level of sea battle model is usually used due to the fact that little effort has been made to maintain the capabilities necessary for performing higher-order analyses. With a few exceptions, notably at Johns Hopkins University Applied Physics Laboratory and in the OPNAV Strategic Nuclear Analysis Branch, Navy battle models are developed on an *ad hoc* basis and are frequently discarded when the developer moves to another job. Although more complex and time-consuming, highly aggregate sea battle models (campaign models) which are usually necessary to perform force level or force mix evaluations have been carried out periodically by specially constituted teams.

Nevertheless, studies based on sea battle models have become a routine part of the Navy force planning and weapon procurement processes. They are a useful and necessary tool to aid in force level and capability deliberations. Senior naval planners must understand (and generally do) that these studies -- particularly the very complex campaign analyses -- are not predictive but indicative in nature, since unpredictable circumstances (including timing, tactical doctrine, or strategic considerations) will heavily influence the actual force engagements. These studies, which synthesize the results of the full hierarchy of sea battle and cost models, have proved to be quite valuable in Navy weapon procurement and force planning decisions.

REFERENCES

Unlike the great majority of the references cited in the remainder of this monograph, all of the studies below are classified. Although they are not available to the general reader, these papers are a highly representative historical record, principally of influential Navy campaign (theater-level) analyses which used micro, medial, and macro models. They are arranged in chronological order.

1. *ASW Force-Level Capability Study (CYCLOPS) (OSD Project 1.I), Office of the Chief of Naval Operations, 16 August 1963.*

2. *A Study of Antisubmarine Warfare Forces for 1970 (CYCLOPS II)*, Center for Naval Analyses, NWG Study No. 31, August 1964.

3. *Exercises for Evaluation of the SSN-593 Class Submarines Project BIG DADDY*, COMSUBDEVGRU TWO, Ser 0183, 20 August 1964.

4. *War at Sea Study (WAS-I)*, Office of the Chief of Naval Operations, Ser 00015P91, 22 April 1966.

5. *Control of the Sea Study (Project CYCLOPS III)*, Office of the Chief of Naval Operations, July 1966.

6. *SUBASWEX 1-67 (LITTLE DADDY II)*, COMSUBLANT, Ser 0021/00293, 23 August 1967.

7. *Navy Fighter Studies I thru IV*, Office of the Chief of Naval Operations, March 1968 through March 1974.

8. *War at Sea II*, Center for Naval Analyses, NWG Study No. 60, August 1968.

9. *U.S. Atlantic Fleet Integrated Escort Tactical Development Program (LANTCORTEX)*, COMASWFORLANT (1971-73).

10. *Naval Requirements and Capabilities, General-Purpose Forces Study (NARAC-G)*, Office of the Chief of Naval Operations, OP-96, July 1971.

11. *Sea Control Exercise (SEACONEX) Program*, COMASWFORLANT (1972-73).

12. *Ship ASW Readiness/Effectiveness Measurement Program (SHAREM)*, Surface Warfare Development Group Program (1972-83).

13. *Sea Control Force Mix Study (SEAMIX-I): An Analysis of the Capabilities of Navy General-Purpose Forces in Sea Control Missions-NATO First Scenario -- 1981*, Office of the Chief of Naval Operations, Systems Analysis Division, Ser 00073P96, 14 August 1972.

14. *Update of the Middle East Power Equation Capability Analysis (MEPEC-73)*, Office of the Chief of Naval Operations, 8 November 1973.

15. *Sea Control Force Mix Follow-On Study (SEAMIX-III)*, *Center for Naval Analyses, NWG Study No. 1044, January 1975.*

16. *Total Allied Forces in the Mediterranean (TALLFORM PHASE II), Center for Naval Analyses, NWG Study No. 1031, April 1975.*

17. *Sea Control Force Mix Study (SEAMIX-II): An Analysis of the Capabilities of Navy General-Purpose Forces, Unilateral Military Action Scenario -- 1981, Office of the Chief of Naval Operations, Ser 96/559696, 6 August 1975.*

18. *Sea Plan 2000, Naval Force Planning Study, Office of the Secretary of the Navy, March 1978.*

19. *Naval Force Mix Study, Convoy (Force Mix/Convoy), Center for Naval Analyses, NWG Study No. 1081, July 1978.*

20. *Sea War 85 -- The Atlantic Campaign, Center for Naval Analyses, NWG Study No. 1139, February 1980.*

ROBERT A. HALLEX

Mr. Hallex has over 20 years' experience as a Naval Operations Analyst, both in the Naval Material Command and on the staff of the Chief of Naval Operations. He has published more than 50 technical studies and analyses of Navy weapon systems and support systems, including annual CNO Program Analysis Memoranda which evaluate total program capabilities in the areas of ASW; AAW; C^3I; Fleet Support/Sealift; Readiness/Sustainability; and Manpower/Personnel/Training. He has been the Study Director or Chief Analyst for a number of total force capabilities studies, including several of those cited in this chapter. His education was in Electrical Engineering and Operations Research. He is also a graduate of the Industrial College of the Armed Forces. His military experience includes 31 years of active and reserve service with the US Air Force, where he accumulated 5,000 hours of operational flying as a pilot in fighter, attack, reconnaissance, and transport aircraft. He is currently Head, Resource/Readiness Appraisal (OP-914), Office of the Chief of Naval Operations.

ABOUT CHAPTER 6

The next two chapters are devoted to what is called support functions, which to the amateur in warfare seem colorless, but to the professional are the invisible foundation for success on the battlefield. In reading these two chapters we urge thoughtful attention to how intimately these models, when properly constituted, are linked to the warfare function. And why not? People and logistics support have long been the submerged part of warfare's iceberg and (however you count) the largest part of peacetime defense expenditures.

In this chapter, Bernard Rostker succeeds in the monumental task of describing the full breadth of model applications to human resource research and management. As he makes clear, human resource models increasingly are being used for manpower, personnel assignment, and training *management* itself. Rostker ably illustrates this from his long personal experience. One of the qualities of human behavior that dominates predictive modeling, he points out, is that people learn and change their ways.

Human Resource Models: An Overview
by
Bernard D. Rostker

INTRODUCTION

A recent Secretary of the Navy frequently reminded his staff that "People are our most important resource." The subject of this chapter is how human resource models help the military departments manage their people so that the United States can meet its defense commitments.

The Overview developed and illustrated some general principles about models and contrasted model types by specific area of application. This chapter illustrates the general principles by highlighting the different types of human resource models, and points out problems that model builders should consider when they develop new models of this type.

In the next pages, we will consider (1) the basic nature of human resource models and what sets them apart from the other models discussed in this volume; (2) a discussion of the various types of human resource models; and (3) a look at the future of human resource modeling.

THE NATURE OF HUMAN RESOURCE MODELS

The reader will note that we use the term "human resource" rather than "manpower." In the military services, the terms "manpower," "personnel," "training," and "assignment" refer to systems that constitute separate organizational entities and management processes by which people are brought together with a weapon or support system in specific jobs to produce something called national defense. Though each system and the models that support it conceptually have their place in an integrated process of human resource management, little has actually been done to develop the interactive models into a single human resource system. Following common usage, we will define the separate systems and later return to the notion of an integrated system for managing human resources. Usually the system is viewed as:

- *Manpower*: the process of determining the numbers and types of people necessary to accomplish a given task.

- *Personnel*: the process of managing people (either directly by management action or indirectly through incentives that affect behavior) so that an appropriate type of person, as defined by the manpower process, is available to be "assigned" to a given unit.

- *Assignment*: the process of matching members of the available pool of personnel to specific jobs.

- *Training*: the process by which a person is given a new set of skills.

Human resource models have characteristics different from the combat models discussed in other chapters. Like logistics models, human resource models tend not to be built to support one-time policy decisions; rather, they have become integral parts of the process of managing human resources. For example, models are used to determine how many people will be recruited and promoted each month, how much they will be paid, and where they will be used. Unlike logistics models, the use of human resource models in personnel management is not new.† Moreover, human resource models focus on the resource that is hardest to model, control, or predict -- people. This places special demands on the modeler, as we shall see.

---

†*A.P. Smith points out that, as early as 1679, the Secretary of the Admiralty regulated the annual entry of officers into the British Navy, and that by 1779, career structures, retention rates, and promotion probabilities were regularly analyzed for the Royal Marines. The systematic collection of the statistics used in personnel planning dates from 1820 in the British Navy, and the basic personnel planning models were discussed in 1899 in the Naval Proceeding of the American Navy. See A.P. Smith, "Defense Manpower Studies,"* <u>*Operational Research Quarterly*</u>*, Vol. 19, No. 3.*

All human resource modelers must determine:

- how to describe people and what type of *aggregation* scheme to use;

- how to take account of the fact that, unlike other resources, people learn and modify their *behavior* in response to changes in both exogenous and endogenous factors.

Aggregation

Any attempt to model a human resource process starts with a decision to describe people according to a well-defined aggregation scheme. The basic building block in a military human resource model is the occupational classification structure. In the Army and Marine Corps, the building block is the Military Occupation Specialty (MOS); in the Air Force, the Air Force Specialty Code (AFSC); and in the Navy, a rating. Regardless of nomenclature, each occupational classification system is an aggregation of tasks into jobs, jobs into positions, positions into occupations, occupations into career fields, and career fields into occupational groups. What should be noted is (1) that each level is a somewhat arbitrary aggregation of an underlying structure; (2) that once a level of aggregation is chosen for a model, variations in the substructure must be ignored; and (3) that the usefulness of a model is directly related to the appropriateness of the aggregation scheme.

There are many examples of human resource models at each level of aggregation. The Services have incorporated highly detailed data about individual tasks as part of the "Instructional Systems Development" process for designing new training programs. The President's Commission on Military Compensation (PCMC), by contrast, used models that aggregated military personnel into cells containing all officers and all enlisted personnel for each year of service; for example, all officers in a given year of service were assumed -- regardless of their background -- to react in the same way to given changes in retirement policy.

A faulty aggregation scheme may lead to unexpected results. The Air Force has assumed that all journeyman avionics mechanics (AFSC 32672A) are homogeneous in their skills and "universally" assignable with equal effectiveness to all models of the F-11, F-15, and F-16. There is some evidence that differences in the tasks that make up the specific jobs on the different aircraft make this assumption invalid.†

The other Services make assignment on more specific classification structures which reflect training and experience on specific equipment. The Navy's Naval Enlisted Classification Code (NEC), for example, is used in conjunction with ratings to match people to specific jobs.

Behavior

The second problem that human resource modelers must face before they start building their models is how to account for people's ability to learn and change their behavior in response to factors incorporated in the model. Many human resource models use historical data on human responses deterministically instead of realizing that the observed responses represent patterns of behavior that will themselves change as endogenous factors in the model change.

This practice is illustrated by personnel planning models that use the transition probability obtained from historical records that an officer will move over time from the nth year-of-service cell to the n+1 year-of-service cell. The model and the historic transition probability are used to explore the effect such policy changes as variations in promotion opportunities and tenure rules have on the officer profile. Yet the very act of changing the tenure rules or promotion opportunities alters not only the policy parameters of the model but also the underlying transition probabilities.†† Failure to account for human behavior compromises seriously the usefulness of the model.

†M.B. Carpenter-Huffman and B.D. Rostker, The Relevance of Training for the Maintenance of Advanced Avionics, 1976.

††A number of economists (Warner, McCall, and Gotz) have estimated the changes in transition probabilities as promotion policies are altered and then have incorporated these "behavioral" relationships into traditional models for personnel planning. See John Warner, Military Compensation and Retention: An Analysis of Alternative Models and a Simulation of a New Retention Model, 1981, and Glenn Gotz and John J. McCall, Estimating Military Personnel Retention Rates: Theory and Statistical Method, 1980.

HUMAN RESOURCE MODELS

Manpower Models

Manpower requirements models are designed to tell the human resource planner how many of what types of people are needed to produce given levels of output. Models range from large-scale simulation types to statistical models that show the numbers and types of people historically used to accomplish a measured amount of work. Simulation models tend to explore manning situations beyond the range of direct observation. By and large, industrial engineering approaches which emphasize statistical analysis of workload and manpower actually employed have gained broad acceptance as the prime tools for determining manpower requirements.† Both types of models suffer from a lack of flexibility (i.e., the ability to substitute one type of labor for another).

Conceptually, *computer simulation models* provide many advantages over traditional industrial engineering approaches. The structure of this simulation model enabled researchers to develop manpower requirements as a function of such elements as flying schedule, component reliability, and frequency of repair and to relate manpower explicitly to projected system output such as sorties flown and ship operations. Unfortunately, simulations are often limited by the poor quality of the basic data.†† SAMSON, developed by the Rand Corporation, is one such case. By at least one account noted in 1968, the quality of data from the Air Force's AFM 66-1 system††† was so poor that it was unable to support the simulations. (Ten years later, when this author had occasion to try to use the same 66-1 data, inconsistencies and incompleteness of the same data frustrated the analysis.)

All too often, simulation models prove the adage "garbage-in/garbage-out." The usefulness of a model may well be limited by inability to obtain reliable data with

---

†*See Richard J. Niehaus, Computer-Assisted Human Resources Planning, 1979.*

††*For a comprehensive review of simulation models in which human performance plays an important part, see A.I. Siegel and J.J. Wolf, Digital Behavioral Simulation -- State-of-the-Art and Implications, 1982.*

†††*Chauncy F. Bell, "SAMSON: A Logistics Simulation," in E.S. Quade and W.I. Boucher, eds., Systems Analysis and Policy Planning, 1968.*

which to estimate its parameters. The analyst needs to give as much attention to the data he "feeds" his model as to the model itself. When someone offers a source of data which reportedly collects everything for all subjects (as the 66-1 data does for the Air Force and 3M data for the Navy), a wise analyst will run and hide. In general, there is nothing so useless as an unreliable census. A carefully controlled and managed sample will almost always provide more reliable and usable information than an undisciplined census.

A more common approach to developing manpower requirements falls under the heading of *industrial engineering*. Commonly, multiple regression statistical techniques are used to establish the relationship between work performed/output produced and the observed level of employment. Data used in the statistical analysis reflect the observed variation in the output and employment levels across a given type of unit, or over time, or both. Here lies a major problem. Having a statistically-derived manpower factor that relates, for example, engine overhauls to manpower required is of only limited use to manpower planners. Planners must also identify the factors that determine the rate that the overhauls will be performed.

Attempts have been made to develop comprehensive planning systems for manpower requirements. The Navy's Manpower Mobilization System (NAMMOS) is a computer-assisted manpower system used by Navy planners and programmers to determine scenario-dependent requirements for mobilization manpower. Industrial engineering workload algorithms, however, account for only 54 percent of the Navy's mobilization manpower requirement. Fixed manning tables account for 28 percent of the manpower, with 18 percent of the total requirement related to policy judgment.

A serious problem with both simulation models and industrial engineering statistical relationships as tools for determining manpower requirements is their general and often implicit assumption that the total manpower requirement is a direct function of the amount of work to be done, no consideration being given to the possibility of combining different types of labor to produce the same level of output at a different -- and often lower -- cost. Important questions of *labor-labor substitutions* are generally ignored.

An important application of economic research to questions of manpower requirements is use of the Constant Elasticity of Substitution (CES) production function and estimates of enlisted personnel productivity for a number of military occupations. Specifically, recent studies show the possibilities of increasing productivity with different mixes of labor. In general, higher-skill occupations tend to overutilize first-termers, and lower-skill occupations tend to underutilize them. Moreover, though the present overall ratio of first-termers to career people falls within the range that would be selected on the basis of economic efficiency, the distributions within specific occupations are not.†

Personnel Models

The traditional feature of military personnel systems is that it has an "in-at-the-bottom, up-through-the-ranks" structure. These systems are frequently modeled as a Markovian process, where movement through the system is determined by a set of transition probabilities.†† These probabilities may derive from either explicit policies -- only a specified percentage of a cohort will be allowed to pass to the next year -- or averages of observed behavior over time. In general application today, personnel planners and managers use three kinds of models: steady-state, dynamic, and transition.

The most common model is the *static or steady-state model*, in which the entire flow system is in equilibrium (i.e., the total number of people who enter the system is equal to the number who leave). Although such models help us understand the long-run effects of policy changes on personnel profiles, they do not show how today's personnel profile will look next year if the transition probabilities are modified directly by policy or indirectly through incentives and behavior.

---

†See Mark J. Albrecht, Labor Substitution in the Military Environment: Implications for Enlisted Force Management, 1979, and Glenn Gotz and C. Robert Roll, Defense Resource Management Study Supporting Papers: The First-Term Career Mix of Enlisted Military Personnel, 1979.

††See Richard C. Grinold and Kneale T. Marshall, Manpower Planning Models, 1977, and J.W. Merck and K. Hall, A Markovian Flow Model: The Analysis of Movement in Large-Scale (Military) Personnel Systems, 1971.

The *dynamic model* takes a given non-equilibrium distribution of personnel and applies a Markovian process to obtain a new personnel profile for a specific period in the future. Thus, though the steady-state/equilibrium solution is not calendar-specific, the results of a dynamic model do change from period to period.

In real-world applications, desirable steady-state policy solutions are often rejected when the near-term effects (as determined by applying the policies through a dynamic model) are shown to be unacceptable. An example is the analysis of alternative retirement systems developed by the President's Commission on Military Compensation. The steady-state analysis of alternative systems showed the benefits of proposed changes in the retirement rules. But applying the new policies to the existing force (as was done in a dynamic analysis) revealed the magnitude of the increases in near-term costs associated with the proposals. The proposals were never adopted.

In recent years, *transition models* have gained in importance. The Army's ELIM-COMPLIP system and the Navy's ADSTAP program are such goal-linear programming models. They incorporate constraints and objective functions (such as manpower requirements and budget ceilings)† and are used to help formulate personnel policies and monitor progress against goals, budgets, and force levels.†† They are designed to describe the force structure that will come closest to satisfying a given set of manpower requirements, given existing personnel constraints. Such applications are useful in the development of "optimal" combinations of policies that deal with a number of real-world constraints simultaneously.

Assignment Models

Manpower models are designed to help personnel planners determine the numbers and types of people needed to perform specific tasks. Personnel models are used to predict the likely outcomes of changes in a personnel policy.

---

†*See A. Charnes and W.W. Cooper, Management Models and Industrial Application of Linear Programming, 1961.*

††*For a brief discussion of the use of ELIM, see Betty W. Holtz and James M. Worth, "The Army's Approach to Improved Strength Forecasting," Defense Management Journal, Vol. 18, No. 2.*

Assignment models, by contrast, are designed to match given individuals with given jobs in such a way as to maximize some objective function, subject to a set of constraints.

As used today, assignment models are employed in the initial classification and assignment of individuals, as well as in their reassignment. Examples of the former are the Air Force's PROMIS/PJM system, the Navy's PRIDE/CLASP system, and the Army's REQUEST/PIM system. These models match men with jobs or school seats. They often consider the attributes of assignees, their preferences, and the goals of the Service.

It should be noted that these matches are not optimal over time. A specific match depends on the time that an individual, billet, or school seat is entered into the model, as well as on previous matches performed. The Army recently proposed a multiyear, multimillion-dollar program to expand the present personnel allocation/assignment system and develop a new one that will more than match recruits against the minimum eligibility requirements for jobs/school seats. Noting the sequential nature of the personnel process, the Army wants to expand its assignment models to be able also to look ahead and answer such questions as these: (a) What is the effect of filling a training seat with a minimally qualified volunteer? (b) What is the "cost" of deliberately leaving a training seat empty? and (c) What is the probability that a more qualified person will become available to fill some specific training seat in the next 24 hours somewhere in the US?

A word of caution for those who develop assignment models: if such models are to be used, they must recognize that their subjects are human, not inanimate, and that people intervene in the assignment process. Many technically elegant models are never used because they are too mechanistic in their approach. Models can help in the assignment process. However, it is not realistic to believe that assignments can be made by a computer without human intervention.

Training Models

The last human resource to consider is training. It can be argued that everything undertaken by the military in preparation for war is training. Though this is generally correct, distinguishing among types of training is useful:

- *Formal training* of individuals usually takes place in a dedicated schoolhouse setting (e.g., basic training and initial skill training).

- *Unit/crew training* sometimes takes place in established training environments but is also part of fleet and unit training exercises.

- *On-the-job training* is incidental to everyday work.

- Formal and informal *skill progression training* is among the requirements for promotion.

The magnitude and diversity of the Department of Defense's training suggest the potential range of models of various kinds. Manpower models help planners determine the numbers of instructors and support personnel needed to train a given number of personnel. Assignment models help make sure that people with a given set of attributes are assigned to specific courses of instruction. Scheduling models help insure the smooth flow of students through the "training pipeline." Simulation models have been developed to help course planners judge the effects of alternative course design, equipment levels, instructor manning, and student flow rates on the operation of the training system.†

FUTURE DEVELOPMENT OF HUMAN RESOURCE MODELS

Earlier, human resource models were considered along the standard lines of manpower, personnel, assignment, and training. Future work will surely refine and extend models in each of these areas. The more innovative modelers, however, are increasingly aware that the functions and activities of the various organizations charged with managing human resources are intimately related. Decisions to change a training curriculum will result in changing work patterns and be reflected in changes in the statements of manpower requirements. Changes in retention patterns affect personnel policies, manpower requirements, and the ability of individual units to conduct on-the-job training.

Although every Service has combined manpower and personnel under a single Deputy Chief of Staff, little has been done to develop an integrated system. The existence

---

†*Margaret B. Carpenter-Huffman, MODIA: Overview of a Tool for Planning the Use of Air Force Training Resources, 1977.*

of such tools as the CES production function and its lack of use by Service planners to explore alternative manpower and personnel structures reflect the continuing fragmentation among the various human resource subsystems.†

A second factor a future model builder should consider is how his model will be used. In any year, many more human resource models are developed than are actually used by the Services in the planning or management of personnel. All too often, the modeler develops an elegant technical model only to find that it is not relevant to the management process he was supposedly trying to affect.

The case of MODIA provides an important lesson. MODIA was designed to help Training Command personnel examine the details of course design and course operation during the planning stage of the process of instruction system development. Though MODIA was a technical success, it was never used by the Air Force. MODIA reflected the desire of the Air Force leadership to search out cost-effective alternatives. It was designed to be used at a local training center during the development of a curriculum, but unfortunately no one -- neither the Air Force leadership nor MODIA's developers -- asked whether those who were to use the new tool had any incentive to search out alternatives. A 1974 field test of MODIA revealed that course planners did not operate in an environment of constrained resources and tended to manipulate MODIA to justify the course designs they favored.

The Air Force leadership which commissioned MODIA and supported its development never understood the incentive structure of those who would use it. The developers of MODIA did not understand how their model would fit into the management structure of the Air Training Command. The

---

†A notable exception is the work by S. Craig Moore to describe and model basic interactions of the Air Force's manpower, personnel, and training systems. See Bruce Armstrong and S. Craig Moore, Air Force Manpower, Personnel, and Training: Roles and Interactions, 1980, and B.E. Armstrong, S.W. Chapel, and S.C. Moore, Air Force Manpower, Personnel, and Training System: Vol. II -- Analysis of the Enlisted Authorization/Assignment and Manpower Requirements Personnel Objectives Subsystems, 1980.

question of implementation was addressed only *after* the model was developed. If adequate attention had been paid earlier, a different -- and one hopes -- more useful product might have been developed.

The experience of MODIA is a good way to end our discussion of human resource models. These models are generally used to manage people. Substantial bureaucratic structures have been built up to perform the personnel, manpower, training, and assignment functions. Models and modelers who support these organizations must not only be skilled technicians but must also understand how a specific model fits into the larger, often implicit, human resource system and must be aware of the incentives of those who will use their models.

REFERENCES

1. Mark J. Albrecht, Labor Substitution in the Military Environment: Implications for Enlisted Force Management, Rand Report R-2330-MRAL, November 1979.

2. Bruce Armstrong and S. Craig Moore, Air Force Manpower, Personnel, and Training: Roles and Interactions, Rand Report R-2429-AF, June 1980.

3. B.E. Armstrong, S.W. Chapel, and S.C. Moore, Air Force Manpower, Personnel, and Training System: Volume II -- Analysis of the Enlisted Authorization/Assignment and Manpower Requirements Personnel Objectives Subsystems, Rand Report N-1476-AF, May 1980.

4. Chauncy F. Bell, "SAMSON: A Logistics Simulation," in E.S. Quade and W.I. Boucher, eds., Systems Analysis and Policy Planning, American Elsevier, New York, 1968.

5. Margaret B. Carpenter-Huffman, MODIA: Overview of a Tool for Planning the Use of Air Force Training Resources, Rand Report R-1700-AF, March 1977.

6. Margaret B. Carpenter-Huffman and Bernard D. Rostker, The Relevance of Training for the Maintenance of Advanced Avionics, Rand Report R-1894-AF, December 1976.

7. A. Charnes and W.W. Cooper, Management Models and Industrial Application of Linear Programming, Wiley, New York, 1961.

8. Glenn Gotz and John J. McCall, Estimating Military Personnel Retention Rates: Theory and Statistical Method, Rand Report R-2542-AF, June 1980.

9. Glenn Gotz and C. Robert Roll, Defense Resource Management Study Supporting Papers: The First-Term Career Mix of Enlisted Military Personnel, February 1979.

10. Richard C. Grinold and Kneale T. Marshall, Manpower Planning Models, North Holland, New York, 1977.

11. Betty W. Holtz and James M. Worth, "The Army's Approach to Improved Strength Forecasting," Defense Management Journal, Volume 18, Number 2.

12. J.W. Merck and K. Hall, A Markovian Flow Model: The Analysis of Movement in Large-Scale (Military) Personnel Systems, Rand Report R-514-PR, February 1971.

13. Richard J. Niehaus, Computer-Assisted Human Resources Planning, Wiley, New York, 1979.

14. A.I. Siegel and J.J. Wolf, Digital Behavioral Simulation -- State-of-the-Art and Implications, American Psychological Services, Inc., Wayne, PA, 1982.

15. A.P. Smith, "Defense Manpower Studies," Operational Research Quarterly, Volume 19, Number 3.

16. John Warner, Military Compensation and Retention: An Analysis of Alternative Models and a Simulation of a New Retention Model, Center for Naval Analyses Report RC-436, August 1981.

DR. BERNARD D. ROSTKER

Dr. Bernard Rostker is an authority in military manpower, personnel, and training. He received his Masters and Doctorate degrees from Syracuse University, where he was awarded a Fellowship by the Department of Labor to complete his doctoral studies on the theoretical and empirical underpinnings of manpower policy. He was introduced to military manpower as a staff economist in the Manpower Requirements Directorate in the Office of the Secretary of Defense (1968-70). From 1970 to 1977 he was a senior economist and Director of "Project Air Force Manpower, Personnel, and Training" at The Rand Corporation. In 1977, he joined the Carter Administration as Principal Deputy Assistant Secretary of the Navy (Manpower and Reserve Affairs) until 1979, when he was appointed Director of Selective Service. In 1981, he left Government and started the Navy Management Program at the Center for Naval Analyses. He joined Systems Research and Applications Corporation (SRA) in 1983, where he currently serves as Director of the Systems Management Division.

ABOUT CHAPTER 7

The professional logistician should read this densely-packed chapter by Stephen Drezner and Richard Hillestad to better understand the shifts of structural emphasis that are taking place (or should be) to tie logistic support models more closely to the battlefield. Even more important, the military man whose knowledge of logistics is shaky should also read this chapter. He will probably then be at once more demanding of the assistance that is possible in the computer age, but more understanding of the huge logistic systems' problems and failures.

Drezner and Hillestad have three important themes. First, the logistics area has been very fertile ground for the development and application of operations research techniques, including mathematical programming, stochastic analysis, and simulation. Second, the reliance on models to help logisticians make management decisions has been increasing as the complexity of support and its effects on warfighting increases. Third, recent indications are that logistics models are evolving toward (1) using more wartime-relevant measures of performance, and (2) more integration with other functions that affect military performance.

Logistics Models: Evolution and Future Trends

by

Stephen M. Drezner and Richard J. Hillestad

I. INTRODUCTION

History and doctrine recognize the key role of logistics in warfighting. However, only since World War II have logistics models been a key element in the planning and operation of support functions. Operations research methods that have been applied to support problems include: linear and dynamic programming; networking and queuing techniques; inventory, reliability, and decision theory; simulations (both man-machine and machine); and statistical models. In many cases these methods were initiated and evolved because of the needs of the military support area.

The number and complexity of weapon systems have grown rapidly since World War II. Modern technology has brought with it large changes in communications, transportation, materials, and data processing. The very nature of warfare has changed in terms of rapidity of attack, severity of weapon effects, increased territorial scale, and complexity of management requirements. Logisticians will rely more and more on models to deal with the complexities of procuring, maintaining, and transporting military material, facilities, and personnel. Models help predict the requirements for support, evaluate the effect of alternative support plans on warfighting capability, and project the readiness of the support system to go to war.

Support modeling has been usefully applied to a wide variety of areas, including:

- Forecasting resource requirements.

- Purchase, stockage, distribution, and data handling for spare parts, ammunition, fuel, food, medical supplies, and construction materials.

- Transportation system design and utilization for peacetime support and strategic mobilization.

- Designing maintainability and reliability into weapon systems.

- Maintenance management policies including inspection, replacement, and workload scheduling.

- Communications, including the design of message and data transmission systems.

- Location and layout of maintenance and repair facilities.

- Personnel and training requirements for maintenance of weapon systems.

This chapter will discuss generally (and incompletely) past and current logistics models in several of these areas and point out that:

1. The logistic support functions have been rich domains for model development and have been at the forefront of certain methodological developments.

2. Despite the sophistication and profusion of many of the methodologies, there are important aspects of military logistics problems that have yet to be modeled adequately.

3. Although there has been some recent progress, much remains to be done with respect to the modeling of logistics support in the context of the real mission of the military -- readiness and sustainability of the operational forces for and in wartime.

II. LOGISTICS MODELS -- THE INTERSECTION OF FUNCTIONAL AREAS, MODELING METHODOLOGY, AND MEASURE FUNCTIONS

Logistics models may be viewed as the intersection of one or more of the various functional areas of support, one or more of the methodologies of operations research, and measurement functions, as illustrated in Figure 9.

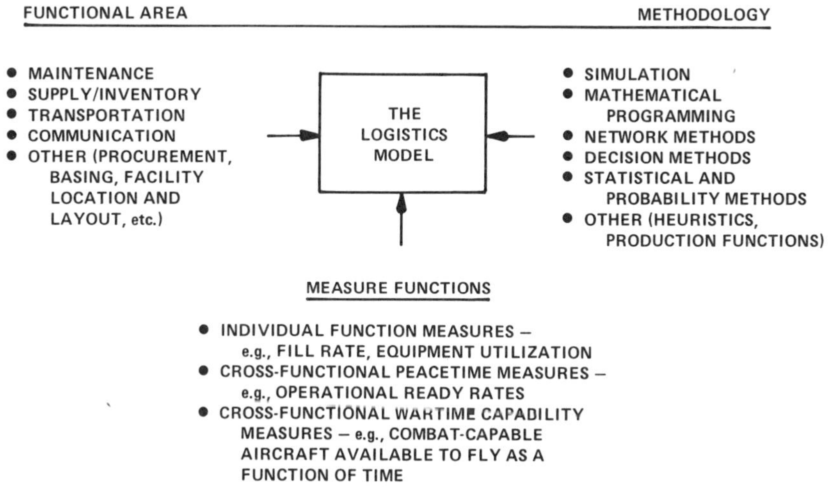

Figure 9. The Logistics Model

In many cases, several different methodologies have been applied to single functional areas. More significantly, there has been a trend toward models that combine several functional areas, permitting much-needed trade-off analysis. The earlier models were driven by measure functions that dealt with how well a particular logistics function was performed. Many of the earlier models used measure functions that focused on internal efficiency rather than system effectiveness. As models began to include several functional areas, the measure function properly focused on more general or system-wide effects. The initial problem with the more general overall objective function was that it usually dealt with system performance in peacetime. Recently there has been movement toward measures that attempt to reflect warfighting capability, but this is still a major area for improvement.

The following illustrations will describe some of the logistic support problems/methodologies and some issues that remain.

MAINTENANCE

Modelers have given considerable attention to several important aspects of maintenance: replacement and inspection policies, reliability, and workload scheduling and planning.

i. Maintenance Replacement and Inspection Strategies

The basic problem for the operator is to determine when to inspect and replace components of weapon systems that are subject to failure. It is desirable to perform this inspection and possible replacement as infrequently as possible because of the high cost of inspection and spare parts and the unavailability of the weapon system during the maintenance period. Letting the system run until failure, however, may be undesirable because of increased cost of maintenance, possible hazardous operation, and unpredictability of failure time (especially during wartime).

Most of the work has focused on simple (non-redundant) system preventive maintenance models, which attempt to reduce downtime of the equipment or minimize maintenance cost to achieve a given level of protection from failure. Generally, these models assume that:

- Time to system failure is unknown, but some aspects of the lifetime probability distribution are known.

- Maintenance regenerates the system. That is, the failure characteristics of the system after maintenance are the same as when the system was new.

- Maintenance actions are independent. Models do not deal with the possible effect one maintenance action has on another or that maintenance often causes maintenance.

To illustrate the varied methodologies applied to these problems, consider that Bellman (1955) applied dynamic programming to determine optimal replacement policies under the above assumptions. Barlow and Hunter (1960) applied a decision-theoretic approach to determine optional preventive maintenance policies; more recently, Arrow, Levhari, and Sheshinski (1972) used a production function approach for the replacement problem. Wagner, Giblis, and Glaser (1964) illustrated the use of mathematical programming for the age replacement decision. A recent and fairly complete survey of models by Sherif and Smith (1981) shows more than 500 articles and books on maintenance replacement and inspection strategies.

Most results describe optimal policies in terms of the nature of the failure distribution and the way it changes in time (increasing failure rate, etc.). A few papers describe procedures for unknown distributions. Cohen (1972) described a process of controlled inspection interval extension for this case. The unknown distribution case seems to be the most useful, as it is rare that failure distributions are known in advance.

A number of aspects of this problem require further development and refinement of the modeling. The objective must include the possible effect of system failure or decreased capability in wartime. This means, among other things, that the nature of the measure function must change and that fluctuating utilization and value over time must be explicitly considered. Also, the failure distributions of damage and wear due to unique wartime use must be considered in addition to normal projections of failure from peacetime utilization.

Most models do not adequately represent the complexity of modern weapon systems. The skill of personnel to diagnose and the ability of test equipment to isolate system failures (in addition to "black box" failures) must all be jointly considered. Additionally, experience with modern weapon systems indicates that maintenance-induced failures and the inability to pinpoint causes of problems produces a heavy workload and causes "removals" that are not the result of any inherent failure characteristic of the component. This raises as a major issue, pervasive across all aspects of maintenance and supply, the usefulness of "mean time between failure" as an indication of demand rather than "mean time between removal." Finally, the interaction of replacement policy and stockage should be considered. All of these factors greatly increase the complexity of the model and constrain the choice of methodologies. For example, closed-form mathematical representations of the process are less likely to incorporate the various interactive, dynamic relationships between system parameters than are heuristic or simulation models.

ii. Reliability

The designer's problem is to design reliable and maintainable weapon systems by using modular components, redundancy, and frequently components with unknown reliability. In most models, reliability is measured in terms of the probability that the system will be operable at a given time; the design problem is to achieve a given level of reliability at minimum cost in components.

Much of the work in this area is stochastic and descriptive, generating basic rules for reliable designs. Barlow and Proschan (1975) present a consistent theory of reliability and life testing in their excellent book. Others, such as Gross and Kamins (1967), have described the application of statistical techniques to determine empirically reliability changes in complex weapon systems. A few papers have described the optimal design of reliable systems using mathematical programming approaches. Using a dynamic programming algorithm, Kettelle (1962) described the least-cost allocation of reliability investment. Jacobson and Arunkumar (1973) applied non-linear mathematical programming to maximize the expected lifetime of series and parallel systems.

This area of maintenance also has considerable need for improved modeling to deal with such issues as the interaction of reliability with supply, maintenance replacement, and measures of wartime capability. A special area of concern is the optimal *improvement* of current system reliability -- that is, the determination of what weapon system components and elements of design should be improved to increase weapon system capability and reduce cost (including supply and munitions costs). Work in this area must also emphasize the wartime use of the system to assess both cost and effect of different levels of reliability.

iii. Workload Scheduling

Scheduling problems or problems of sequencing abound in all aspects of military planning. In the maintenance function, special attention has been given both to the problem of scheduling work as it arrives in a shop (workload scheduling) and the problem of planning the sequencing of work requiring multiple tasks and equipment. The well-known shop flow problem has N jobs of various duration for M machines; procedure requirements for processing; and usually a flow time, due date, or machine utilization objective. This simplified representation is a very difficult combinatorial problem which many modelers have studied with only limited success. Generally, only the two-machine problem has a known solution, although results for more complicated problems have been stated. Conway, Maxwell, and Miller (1967) characterize the shop flow scheduling problem and a wide variety of possible solution approaches. Maxwell (1969) applied simulation to assess various types of rules, and many articles (such as an early one by Johnson (1954)) describe rules of sequence.

Workload scheduling increases the complexity of the model by including the random generation of demands for processing. In military application, this randomness is associated with the random failure of weapon system components as they are used. Most of the successful modeling for this problem has been with simulation (see Miller (1973), for example) and special network techniques such as GERT (Pritzker and Hopp (1966)) and Q-GERT (Pritzker (1977)).

A criticism that applies to almost all models developed for this problem is that the flow time, utilization, and due date objectives are inadequate measures, particularly when the scheduling is for components of weapon systems whose ultimate purpose is warfighting. In general, deterministic N-job, M-machine models have little relevancy to military maintenance. Military models need to examine decisions about priority repair and expedited repair in the face of stochastic demands. There may be N jobs but seldom more than one or two machines in military models. An avionics repair shop is a good example. The workstations (machines) are sufficiently specialized so there are only a few that can perform the same job. The degree of mission essentiality and criticality of the components is rarely included. For example, when adequate spare components exist, the need for immediate repair of an item may not be as high as for an item with no spares. A component that will degrade mission capability only slightly will have less priority than one with high essentiality. Priority repair, a necessary aspect of workload scheduling in the military, is largely ignored in current scheduling models. While these weaknesses in current models do not make them totally useless to the operator or decision maker, they certainly limit their value.

SUPPLY AND INVENTORY ANALYSIS

Multi-echelon inventory analysis has dominated the modeling work in this area. The problem is to determine ordering, inventory, and distribution policies for two or more interrelated supply or production facilities. In the military these represent depot repair facilities, intermediate repair shops, and the inventories associated with each. Although the basic effect of inventory policies is on availability of weapon systems, more intermediate measures -- such as the probability of satisfying demands for components and the expected number of components backordered -- have generally been used. Most inventory models make assumptions about the nature of demands -- whether they are deterministic or stochastic, whether they are stationary or non-stationary, whether demands not satisfied immediately are lost, and whether demands arise from a single type of component or from multiple types. The inventory review process is assumed to be either continuous or periodic, and the components may or may not be repairable.

The spare parts problem and maintenance problems are closely related. Sufficient spare parts can allow weapon systems to operate while components are being repaired. Flexible priority in maintenance can quickly repair weapon systems when spares are short. Generally, multi-echelon inventory models consider the maintenance and transportation of components (the converse is not usually true for maintenance models), although the information used (average repair time, location of repair, and average transportation time) is usually minimal.

The multi-echelon problem is of extreme interest to the military because of the influence component shortages have on weapon system capability and because the majority of logistics resources (manpower, depots, test equipment, transportation, etc.) are included in it.

Several authors have developed models for this problem:

- Zangwill (1966) described a deterministic approach and characterizations of solutions for a multi-product, multi-facility production and inventory problem.

- Veinot (1969) recognized the network interpretation of this problem.

- Sherbrooke (1968) applied marginal analysis and the use of Lagrange multipliers to determine optimal supply policies.

- Muckstadt (1973) extended the Sherbrooke work for indentured components (subcomponents of components).

- Kotkin (1978) developed an heuristic algorithm for the multi-echelon indentured item supply problem.

- Hillestad and Carrillo (1980) extended the Sherbrooke and Muckstadt work for non-stationary demand processes.

- Clark and Scarf (1960) applied dynamic programming to the periodic review multi-echelon problem.

Another aspect of the supply/inventory problem includes the distribution of inventory to meet the time-dependent and mobility requirements of the military. Miller (1968) used simulation to investigate alternative real-time distribution rules for components shipped from depots to air bases. Landi (1967) used mathematical programming to determine optimal prepositioning rules for spare parts for military transport aircraft.

Multi-echelon inventory models typically treat the sources of repair and capability to repair as known elements of the problem. The Level of Repair Analysis (LORA) problem complicates the models by letting these be decision variables as well. Fixed costs of facilities manpower, test equipment, transportation links, etc., frequently lead to large integer models. (This area is further complicated by the politics surrounding the location of military facilities and the commonality of facilities, equipment, and skills for some new and old weapon systems.)

The major criticism of work in the inventory/supply area for military applications is the lack of attention to objectives that emphasize weapon systems availability and capability. The use of a "backorder" measure does not consider the relative importance of different components of weapon systems and does not properly reflect their availability even when all elements are essential; nor does it account for the cross-cannibalization (consolidation of shortages on the smallest number of weapon systems) that commonly occurs. Other important features of actual logistics operations -- such as lateral supply (borrowing components from other bases); priority repair of backordered, critical components; and priority transportation from military depots -- have not been treated in most models. Wartime military operations with uncertain arrival of transportation and possible disruption of repair are ignored in most of these models as well. Most of the modeling assumes that something is known about the demand for maintenance and spare parts in terms of its expected value or probability distribution. In truth, we know very little about the wartime failure and removal patterns of components from new weapon systems. Inventory/supply models should be designed to reflect this uncertainty when generating optimal stockage policies. Most current models focus on peacetime optimization issues. Their answers can be quite misleading when applied to wartime situations. Modeling the wartime environment is, of course, more difficult, but progress is being made.

TRANSPORTATION

The basic objective is to design a network of routes or carriers to move the materials and personnel of war. The overriding aspect of this problem is to create a transportation system that can respond in a timely manner to varied mobilization requirements. The cost of creating such a transportation system and maintaining it in peacetime brings in economic considerations and the trade-off between time and network capacity. The capacity of military transportation is generally described in ton miles per day -- that is, how many tons can be moved how many miles per day. Why this needs to be changed is discussed below.

Transportation problems during and after World War II sparked the development of that special class of network algorithms called transportation algorithms. Dantzig (1962) and Ford and Fulkerson (1954) provided early developments in these algorithms. The special case with the military time objective was called a "bottleneck" transportation problem, and Hammer (1969) and Garfinkel and Rao (1971) developed special algorithms for it. Fulkerson and Dantzig (1954) used integer mathematical programming to include additional real constraints in tanker scheduling. Others resorted to simulation. For example, Nolan and Mastroberti (1972) simulated an entire airlift operation to understand the quantitative need for C-5A transport aircraft, and the Defense Department currently uses a large simulation called MIDAS to study airlift requirements and capabilities.

The use of ton miles per day as a measure of the military transportation system is inadequate for many purposes because it does not state the effect of shortfalls in transportation capacity or the effect of delays on output measures of the total support system, not just the transportation system. Since transportation is generally a key element of weapon systems support (transporting those systems, spare components for those systems, and maintenance personnel), it should be measured in terms of the capabilities of those systems. Models must consider the effect of losses and degradation of transportation on the operation of those systems. Furthermore, the value of priority transportation for certain commodities in a multi-mode, multi-commodity transportation system for wartime support must be measured in terms of the effect on each weapon system at each location. An example of a model that attempts to relate the performance of the transportation function to relevant system-wide output measures is discussed by Berman et al. (1981). The interaction between transportation and other logistics functions and their impact on sortie generation capability over time is dealt with explicitly. Trade-offs between more spares and more transportation or more transportation versus more weapon systems can be performed directly by the Berman Model. This study is an excellent example of useful analysis of the value of transportation to wartime capability.

COMMUNICATION

Most modeling work in the communication area has been done on the synthesis problem -- designing communication trunks and switching networks for given capacities at minimum costs. Problems of special interest in military communications include message criticality, high reliability requirements, the potential loss of trunks and switching centers, and message security. Kalaba and Juncosa (1956) described a linear programming approach to synthesis and message routing and suggested the use of weighting functions for critical messages. Gomory and Hu (1964) recognized the applicability of various network approaches to the problem. Cady, Lientz, and Willsworth (1974) described experiments in communication networks to determine the trade-off of economic efficiency through centralization and vulnerability.

OTHER FUNCTIONAL AREAS

The logistics areas described above represent only a fraction of the military problems for which operations research models are needed and have been usefully applied. Facilities location and facilities layout for effective maintenance and reducing vulnerability of weapon systems have been addressed with varied mathematical programming algorithms. Various forms of centralization or decentralization problems in supply, maintenance, and other support functions have received attention over the years. There are economic incentives to centralize, and these tend to dominate if a peacetime environment is driving the design. During wartime, span of control and vulnerability issues tend to favor various forms of decentralization. The issues are difficult to represent in a model, and there does not seem to be an easy general answer. The appropriate balance appears to be very specific to the scenarios of interest. The methodology to study these trade-offs does exist.

Procurement problems in terms of bid evaluation, phased procurement of support materials, facilities expansion, planning, and budgeting are important problems requiring and frequently using the models of operations research. We do not, however, discuss these in this chapter.

Clearly, the functional areas of logistics support have been fruitful ground for the development of modeling methodologies, and many models have contributed to the effectiveness of the logistics functions. Still, as we have briefly indicated, much remains to be done to expand the scope and realism of those models within the functional applications; and, of course, there is the continuing desire to get more integration of models between functional areas. Frequently, models ignore military objectives and constraints in favor of peacetime economic objectives or intermediate measures of the particular function being examined. The next section gives some historical perspective on this issue and discusses the evolution that models are undergoing for a more direct consideration of the influence of logistics on wartime capability.

III. HISTORICAL TRENDS AND THE NEED TO REFOCUS THE LOGISTICS MODELING EMPHASIS AND SCOPE

MODELING EMPHASIS AND SCOPE

Since the end of World War II and up to the early 1970s, the preeminent emphasis on defense was strategic deterrence. The de-emphasis of tactical conventional capability because of the nuclear umbrella caused much of the logistics concern to be directed to the peacetime support of forces. The long period of peacetime activity for the military, the reduction in real defense appropriations, and the increasing cost of sophisticated weapon systems motivated logistics modelers to look for cost minimization within peacetime performance constraints. Logistics, becoming increasingly costly and yet loosely related to operational performance (most models, as illustrated in the previous section, used intermediate objectives associated with the specific functional areas), was highly vulnerable to military budget cutting, even within the defense establishment. The effect on warfighting capability of $100M fewer spares was simply not apparent. The effect of not having 10 aircraft or 100 tanks was more intuitively obvious.

The result of this economic emphasis and lack of distinct relationships to operational performance led to models focused on peacetime efficiency. Structures and policies were based on least-cost peacetime alternatives. Arguments for centralization and decentralization of maintenance and supply were generally based on peacetime economic considerations rather than potential wartime support requirements.

Visions of possible future wars include: a very dynamic environment, entirely mobile threats, potential for high vulnerability, and generally a great deal of uncertainty. Wars may be fought anywhere on the globe. Nuclear weapons may be used in sustained operations. War may be just conventional or combined conventional and nuclear. The objectives, constraints, and structures of logistic support models must now deal more directly with the dynamics, uncertainty, and mission objectives of warfare.

Clearly, support requirements are likely to be more dynamic in wartime than in peacetime. The failure rates of weapon systems, availability of maintenance during a deployment, and surges in weapon systems activity rates will place different requirements on the support structure than will the relatively constant peacetime activity rates. Muckstadt (1980) showed, for example, that the spare parts requirements for a surge of activity are incorrectly approximated with the stationary inventory models currently in use. Lippiatt et al. (1981) showed that priority repair would clear a bottleneck on an aircraft carrier, whereas approximations with a steady-state queuing result predicted an infinite queue. In that case, the question was not whether the maintenance facility was overloaded but how quickly the overloading became apparent in terms of weapon system performance under wartime activity rates.

New logistics models must consider the uncertainty of tactical wartime support requirements. The models must provide robust answers to support requirements when the scope, scale, and type of operational activity is only an educated guess. Uncertainty regarding losses and damage to support systems and about weapon system characteristics such as failure rates, mission effectiveness, and wartime maintenance requirements are key issues. The support environment itself will be more chaotic in wartime. (Modern military supply systems are almost totally dependent on large centralized computer systems with supporting computer systems at each base. Capability after mobilization -- when computer support may be lost -- could be seriously degraded.)

Support objectives in wartime are different from those in peacetime. Efficiency, quality-of-life issues, and costs will take a back seat to mission effectiveness. Models tuned to wartime logistics support will recognize that support is provided for maximum effectiveness of the weapon systems and forces supported.

Finally, because future tactical warfare may be considerably different from previous wars, models based on past views of wartime support requirements should be rethought. More capable weapons (in payload and accuracy) and increasing vulnerability of support make logistics disruption an important consideration. A great ability to inflict damage may mean a short intense conflict, with an increase in the requirements for carefully calculated initial support. Furthermore, the sophisticated tactical weapon systems currently planned and in use require different kinds of support than past weapon systems in terms of complex test equipment, skilled technicians, and high reliance on communications and computers.

Logistics models are beginning to accommodate these new views. A few recent examples include: the work on dynamic multi-echelon inventory systems by Muckstadt (1980) and Hillestad (1982); the evaluation of support in a theater of war under battle damage conditions provided by Emerson's TSAR-TSARINA models (1982); the Berman et al. (1981) use of wartime models to investigate theater transportation systems; and the US Air Force's development of War Reserve Spares Kits based on measures of aircraft availability rather than supply backorders.

IV. SOME IMPORTANT REMAINING ISSUES IN LOGISTICS MODELING

Military modeling areas are merging and should continue to do so. For example, battle management models need to consider logistics support (particularly in the case of damage and losses). Weapon procurement models must become highly sensitive to the support requirements. (Requirements for sensitive, complex test equipment may be inconsistent with the probable environment of the weapon system.) Manpower models need to show the influence of manpower on the supportability of weapons, and trade-offs between personnel skills and design of complex weapon systems need to be understood.

The drive for more operationally-oriented performance measures for support models must continue. Weapon systems availability is a start, but it is certainly desirable to consider mission capability, mission effectiveness, etc., as well.

There should be more effective use of good support models in the ongoing management of logistics. We must move from the bean-counting approach of current readiness reporting to the use of models to predict force capability to go to war as it is affected by current logistics states. Implementation issues and the management structure affecting model use continue to be important.

V. CONCLUDING REMARKS

We have presented only a sample of the functional areas and logistic support models, but it should be clear that logistics has been and will continue to be rich territory for development of operations research models. Mathematical programming, stochastic analysis, simulation, and other approaches have certainly made important contributions to military logistics. New models are evolving toward the wartime support aspects of logistics and increasingly relevant measures of military performance. Finally, as weapon systems and support requirements become more sophisticated and complex, there are important issues requiring more integration of military modeling areas. Examples of the wider application of these newer models include: studying trade-offs between short-term and long-term goals (e.g., between readiness and force modernization) and studying the preferred allocation of relatively fixed dollars among competing but interrelated demands (i.e., people, test equipment, spares, and transportation).

REFERENCES

1. K. Arrow, D. Levhari, and E. Sheshinski, "A Production Function for the Replacement Problem," The Review of Economic Studies, vol. 39(3), no. 119, pp. 241-249, July 1972.

2. R.E. Barlow and L.C. Hunter, "Optimum Preventive Maintenance Policy," Operations Research, vol. 8, no. 1, pp. 90-100, 1960.

3. R.E. Barlow and F. Proschan, Statistical Theory of Reliability and Life Testing Probability Models, Holt, Reinhart, and Winston, Inc., New York, 1975.

4. R. Bellman, "Equipment Replacement Policy," Journal of the Society for Industrial and Applied Mathematics, vol. 3, no. 3, pp. 133-136, 1955.

5. M.B. Berman, J.M. Halliday, M.J. Carrillo, and N.Y. Moore, Combat Benefits of a Responsive Logistics Transportation System for the European Theater, The Rand Corporation, R-2860-AF, December 1981.

6. G.M. Cady, B.P. Lientz, and N.E. Willsworth, "Experiments in Communications Networks," Navy Research Logistics Quarterly, vol. 21, no. 1, pp. 107-124, March 1974.

7. A.J. Clark and H. Scarf, "Optimal Policies for a Multi-Echelon Inventory Problem," Management Science, vol. 6, no. 3, pp. 475-490, April 1960.

8. I.K. Cohen, Aircraft Planned Inspection Policies: A Briefing, The Rand Corporation, R-1025-PR, June 1972.

9. R.W. Conway, W.L. Maxwell, and L.W. Miller, Theory of Scheduling, Addison-Wesley Publishing Co., Massachusetts, 1967.

10. G.B. Dantzig, Linear Programming and Extensions, Princeton University Press, New Jersey, 1962.

11. D.E. Emerson, An Introduction to the TSAR Simulation Program: Model Features and Logic, The Rand Corporation, R-2584-AF, February 1982.

12. L.R. Ford, Jr. and D.R. Fulkerson, Maximal Flow Through a Network, The Rand Corporation, RM-1400-PR, November 1954.

13. D.R. Fulkerson and G.B. Dantzig, "Minimizing the Number of Tankers to Meet a Fixed Schedule," Naval Research Logistics Quarterly, vol. 1, no. 3, pp. 217-222, 1954.

14. R.S. Garfinkel and M.R. Rao, "The Bottleneck Transportation Problem," Naval Research Logistics Quarterly, vol. 18, no. 4, pp. 465-472, December 1971.

15. R.E. Gomory and T.C. Hu, "Synthesis of a Communication Network," Journal of the Society for Industrial and Applied Mathematics, vol. 12, pp. 348-369, 1964.

16. A.J. Gross and M. Kamins, Reliability Assessment in the Presence of Reliability Growth, The Rand Corporation, RM-5346-PR, September 1967.

17. P.L. Hammer, "Time-Minimizing Transportation Problems," Naval Research Logistics Quarterly, vol. 16, no. 3, pp. 345-357, 1969.

18. R.J. Hillestad, Dyna-METRIC: Dynamic Multi-Echelon Technique for Recoverable Item Control, The Rand Corporation, R-2785-AF, March 1982.

19. R.J. Hillestad and M.J. Carrillo, *Models and Techniques for Recoverable Item Stockage When Demand and the Repair Process Are Non-Stationary -- Part I: Performance Measurement*, The Rand Corporation, N-1482-AF, May 1980.

20. S.E. Jacobson and S. Arunkumar, "Investment in Series and Parallel Systems to Maximize Expected Life," *Management Science*, vol. 19, no. 9, pp. 1023-1028, 1973.

21. S.M. Johnson, "Optimal 2 and 3 Stage Production Schedules with Setup Times Included," *Naval Research Logistics Quarterly*, vol. 1, no. 1, pp. 61-68, 1954.

22. R.E. Kalaba and M.L. Juncosa, "Optimal Design and Utilization of Communication Networks," *Management Science*, vol. 3, no. 1, pp. 33-44, October 1956.

23. J.Q. Kettelle, Jr., "Least-Cost Allocation of Reliability Investment," *Operations Research*, vol. 10, no. 2, pp. 249-265, March-April 1962.

24. M. Kotkin, "An Heuristic for Multi-Echelon, Multi-Indentured Inventory Problems," *AD Technical Report TR 79-1*, U.S. Army Inventory Research Office, December 1978.

25. D.M. Landi, *Positioning Recoverable Spares in Military Airlift Networks*, The Rand Corporation, RM-5087-PR, March 1967.

26. T.F. Lippiatt, R.J. Hillestad, L.B. Embry, and J. Schank, *Carrier-Based Air Logistics Study: Integrated Summary*, The Rand Corporation, R-2853-NAVY, December 1981.

27. W.L. Maxwell, *Priority Dispatching and Assembly Operations in a Job Shop*, The Rand Corporation, RM-5370-PR, October 1969.

28. B.L. Miller, *A Real-Time Metric for the Distribution of Serviceable Assets*, The Rand Corporation, RM-5687-PR, October 1968.

29. L.W. Miller, *VIMCOS II: A Workload Control Simulator Model for Exploring Man-Machine Roles in Decision Making*, The Rand Corporation, R-1094-PR, June 1973.

30. J.A. Muckstadt, _Comparative Adequacy of Steady-State Versus Dynamic Models for Calculating Stockage Requirements_, The Rand Corporation, R-2636-AF, November 1980.

31. J.A. Muckstadt, "A Model for a Multi-Item, Multi-Echelon Inventory System," _Management Science_, vol. 20, no. 4, pp. 472-481, 1973.

32. R.L. Nolan and R. Mastroberti, "Productivity Estimates of the Strategic Airlift System by the Use of Simulation," _Naval Research Logistics Quarterly_, vol. 9, no. 4, pp. 737-752, 1972.

33. A.A.B. Pritzker, _Modeling and Analysis Using Q-GERT Networks_, Halsted Press and Pritzker Associates, Inc., 1977.

34. A.A.B. Pritzker and W.W. Hopp, "GERT: Graphical Evaluation and Review Techniques," _Journal of Industrial Engineering_, vol. 17, no. 5, 1966.

35. C.C. Sherbrooke, "METRIC: A Multi-Echelon Technique for Recoverable Item Control," _Operations Research_, vol. 16, no. 1, pp. 122-141, January-February 1968.

36. Y.S. Sherif and M.L. Smith, "Optimal Maintenance Models for Systems Subject to Failure -- A Review," _Naval Research Logistics Quarterly_, vol. 28, no. 1, pp. 47-74, 1981.

37. A.F. Veinot, Jr., "Minimum Concave Cost Solutions of Leontief Substitutability Models of Multi-Facility Inventory Systems," _Operations Research_, vol. 17, no. 2, pp. 262-291, March-April 1969.

38. H.M. Wagner, R.J. Giblis, and R.G. Glaser, "Preventive Maintenance Scheduling by Mathematical Programming," _Management Science_, vol. 10, no. 2, pp. 315-334, 1964.

39. W.I. Zangwill, "A Deterministic Multi-Product, Multi-Facility Production and Inventory Model," _Operations Research_, vol. 14, no. 3, pp. 486-507, May-June 1966.

DR. STEPHEN M. DREZNER

Dr. Drezner is a Vice-President of The Rand Corporation, responsible for their work for the United States Air Force. Former activities in military operations research were primarily concerned with readiness measurement; logistics systems; and manpower, personnel, and training. He received his B.S. in Mathematics from Queens College and a B.S. in Industrial Engineering from Columbia University. He has an M.S. in Quantitative Analysis from the University of Southern California, where he also completed coursework leading to a Ph.D. in Quantitative Analysis.

DR. RICHARD J. HILLESTAD

Dr. Hillestad is a Senior Engineer with The Rand Corporation. He has focused his work on the application of management science and computer science techniques to the solution of military problems in logistics, command and control, and manpower. He received his B.S. and M.S. in Electrical Engineering from the University of Wisconsin and was awarded a Ph.D. in Engineering and Applied Science by the University of California at Los Angeles.

ABOUT CHAPTER 8

The Overview emphasizes that when models are used to shed light on defense policy and programming decisions, then certain properties, among them flexibility and transparency, are particularly important because decisions made today affect the remote future, which will be inadequately predicted at best and may follow many paths. The next three chapters illustrate the difficulties, some successes, and some rude failures of modeling for policy and procurement applications.

In Chapter 8, Englund discusses modeling to help select the best weapon system from among competing possibilities. Weapon cost and effectiveness data are combined in a trade-off context to provide information a decision maker can use to select initial system characteristics *early* in the development process.

At the outset of this chapter, Englund enumerates seven imperatives of successful weapon system procurement modeling. He then illustrates from his own experience, principally in Air Force-related decision making, how those imperatives apply in practice, and how neglect of them led into some analytic *cul-de-sacs*.

In brief, the procurement process wants models that are simple, transparent, and flexible to allow many variations in order to adapt to the many uncertainties that are inescapable when making weapon choices ten years or more in advance of system operation and to allow the decision maker to understand what drives weapon performance and costs. Such models tend to be specially tailored, *ad hoc*, and not generally applicable to other problems.

Weapon Procurement Modeling

by

John A. Englund

In this chapter we deal not with models but rather with the modeling process. The models for weapon procurement analysis have tended to be special-purpose (called *ad hoc* in the Overview) rather than for general use and most often not of a size or complexity that merits extensive documentation and broad dissemination. Thus, rather than discuss specific models, we make seven points that are important for anyone doing weapon procurement analysis†:

1. Modeling for weapon selection is essential because it has become too expensive and time-consuming to build and test prototypes.

2. The primary criteria for weapon selection are comparative cost and effectiveness.

3. Techniques used to rank candidates for weapon procurement usually cannot be carried over into force-level analysis.

4. The many uncertainties are of major importance since the weapons are often conjectural (paper designs) and will operate in a far-future environment.

5. Simplicity is especially valuable because of the many uncertainties. The model must have flexibility and adaptability for many effectiveness and cost excursions.

---

†These are "old hat" to the experienced modeler, but also (in our experience) too often ignored, even by the experienced; thus, they merit repetition.

6. The modeling must allow (perhaps even require) new employment doctrine.

7. For maximum utility, the modeling should be done in close working relationship with the client. Since there is no one "answer," the model should shed light on the trade-offs for the decision maker.

Weapon procurement is linked to systems analysis: the selection, by some appropriate set of decision criteria, of a group of characteristics for a new weapon system; the comparison of alternative systems; or the choice of technological improvements to a system. Two major changes from battle analysis affect the choice of model. First, of course, there must be an estimate of the procurement and operating costs of the candidates.† Second, the modeler is dealing with the characteristics and performance of new (and therefore conjectural) weapons. Weapon procurement analysis has been made necessary by the increase, over the past several decades, in weapon costs and procurement times.†† It has been made practical in proper depth and detail by advances in computer hardware and software capabilities.

Analytic techniques for cost-effectiveness comparisons are many and varied. The details are outside the scope of a short paper on modeling, but a brief comment is in order. A common procedure is to first establish equal life cycle cost sets of competing weapon systems and then to compare their effectiveness, using models pertinent to different employment scenarios. It is less common, and may be bureaucratically less acceptable, to establish the numbers of equally effective weapons and then estimate life cycle costs. Very often the "best" model to use in a weapon selection problem, particularly in the conceptual stage of weapon development, is of the diagnostic or exploratory type.

What makes weapon procurement analysis possible is the power of cost and effectiveness *comparisons*. If model results show that a candidate fighter will shoot down six

---

†*The reader who believes that cost modeling is easy is referred to Gene Fisher's work (1975) for an introduction to the pitfalls and techniques involved.*

††*See Quade and Boucher (1968) for an excellent discussion.*

enemy fighters during its combat lifetime under circumstances in which an existing fighter shoots down only two, then while the analyst and his client may not believe that either six kills or two kills will hold true in wartime, they can put considerably more trust in the 3:1 improvement. At the same time, if under comparable and reasonable assumptions the new fighter costs several times as much as the old fighter, buying or retaining three times as many of the old fighters may be the preferable alternative.

Other techniques for making performance and cost comparisons are available to the analyst. For example, the Navy's *Major Fleet Escort Study* (1967) examined the characteristics of alternative surface combatants in a fixed 90-day war scenario using fixed 10-year lifetime systems costs. Neither assumption was taken literally, but enough variations were explored to convince the Navy and OSD clients that the results were not sensitive to the assumptions.

Since such techniques will probably be used to rank candidates, it is a chancy business to transfer results of weapon selection work directly over into force-level work -- the question of how many of the preferred alternative to buy. For the former, it is reasonable to assume that the fighters are handled competently; for the latter, one has to worry about how many aircraft are destroyed on the ground, have no fuel, or may be flown by poorly trained pilots.

At the same time, the analyst interested in weapon procurement must often infringe on the domain of the force planner. An important part of the selection of system characteristics may well be the question of how many might be put into inventory, particularly when a new system is being studied as the replacement for a large and aging force of some kind. For example, in the early days of the B-1 (*née* AMSA -- Advanced Manned Strategic Aircraft) program there was an important trade-off between number of aircraft and aircraft payload. This was because the new aircraft was intended to replace (on at least an equal capability basis) a specified number of B-52s, and the cost-effectiveness comparisons led through several stages of increasing payload (and therefore increasing aircraft weight and cost).

And this just begins to broach the subject of analytic techniques to estimate relative desirability. The caution we invoke is not to allow data developed for choosing between weapons to be heedlessly used to determine how many of the weapons to purchase.

The modeler must struggle with many uncertainties. If the model (scenario) is too abstract, it may be meaningful only to the modeler, and therefore impugned. If it is too detailed, it carries the liabilities that all detailed battle models have and, in addition, may lead to testing a future system against the enemy's present weapons and tactics. Inevitably there is debate over the characteristics and tactics of the future opposition, the friendly force mix, the costs attributable to the new weapon, and the geography. There are additional uncertainties: the design of the weapon (bomb, missile, aircraft, submarine,...) will doubtless include technology that has not yet been fully developed and that may then not perform in accordance with the assumptions; the analyst-perceived mission for the weapon may never be realized; the weapon's vulnerability to enemy weapons is unknown; the environment in which the weapon will be used, perhaps 10 or 20 years hence, can be only dimly seen; the threats that will face the weapon may be developed only as a response to its development; and cost-data development presents a major challenge.†

The modeler, of course, does not work in isolation. Scenarios should be developed with inputs and active participation on the part of planners, operators, and intelligence and cost personnel.

Although not universally the case, weapon selection models tend to be relatively simple, transparent, deterministic (because the inherent uncertainties in many weapon selection problems do not justify attempts at "greater precision"), and analytic (often using a computer for the drudgery). They should, ideally, produce tradeoffs between weapon characteristics or between weapons designed to do the same job. Appropriate decision criteria may (and often do) change through the evolving analysis of a new concept, particularly as the decision level escalates. It is impossible to anticipate all of the

---

†*For an excellent study of these challenges in the case of the TFX, see R.F. Coulam, Illusions of Choice: The F-111 and the Problem of Weapons Acquisition Reform, 1977.*

eventual changes in developing a single model that can handle all variations. Thus, the original model will most likely undergo significant overhaul or be discarded entirely and replaced -- a strong argument for model austerity and flexibility.

For example, some years ago a technologist-level analysis of a proposed improvement to the Minuteman III system used the number of warheads per missile as the measure of merit. The proposed increase from three to four was rated as a one-third improvement in force effectiveness, ignoring the facts that not all Minutemen were IIIs and not all IIIs would carry the maximum possible number of warheads. As the decision whether to proceed with the modification moved to higher levels, criteria such as damage effectiveness per missile and number of warheads deliverable by the ICBM force came to be applied (and, we would add, the apparent cost-effectiveness of the modification decreased).

As a second example, an electronic warfare device may initially be justified in the context of a wide range of other means of protecting the aircraft. However, defensive measures for aircraft protection quite often cannot be considered in isolation from offensive uses of the aircraft because of the impact on the mission for which the system is intended to be used. A new "black box" requires volume, weight, and power from the total available in the aircraft. Too often these come at the expense of range, combat time, or offensive payload, so their impact on the mission of the system must be considered among the decision criteria.

Because of the uncertainties, the modeler is tempted to use some form of optimization model, usually involving some aspects of zero-sum, two-person games.† Sadly enough, optimization usually leads to good theoretical results, but too often to poor practical ones (e.g., defending only high-value targets when the experienced military planner would defend all targets to some extent; or allocating essentially all offensive resources to attack only a few high-value targets, again contrary to accepted practice, except perhaps in scenarios that are severely weapon-constrained). The point, of course, is

†*For example, John Danskin, The Theory of Max-Min and Its Application to Weapons Allocation Problems, 1967.*

that though the model be simple the results must make sense in a practical military environment.

Sometimes the R&D planner, in his rush to "sell" his new concept as a viable one and establish a place for it in the budget, overlooks seemingly unimportant items, such as improved tactics, that would improve the effectiveness of one candidate more than another. He may do an excellent job of developing the concept to take good advantage of developing technology but may overlook the fact that the new system must also be cost-effective in competition with the old system.

For example, early concepts of a new strategic aircraft envisioned an employment doctrine modeled on older, bomb-dropping aircraft, neglecting the advantage associated with new-technology, air-to-surface missiles that both the new and the old aircraft could carry. This put a "cap" on the payload (number of weapons) to be carried by the aircraft, essentially determined by crew ability to handle the many details associated with "visiting" each of x targets. Insufficient advantage was taken of developing computer technology and the benefits of "launch and leave" missiles. The resultant aircraft "competed" very poorly with existing systems, and that situation did not change until the analyst could convince the planner that the solution was increased and improved payload -- hence a larger and costlier aircraft, but one with decreasing marginal costs for additional weapons.

Great burdens are placed on the modeler because clients are interested in a lot of detail about the effects that small changes in characteristics -- 10 percent changes in speed or range or delivery accuracy -- may have on performance in combat. It is difficult to isolate the effects of a parameter change for different circumstances (e.g., an interceptor role versus a fighter escort role) without masking some side effect -- often adverse. The modeler is searching for ways to capture non-linearities, breakpoints at which effectiveness takes off or drops off a cliff, or leverages in which a small change of one characteristic compounds the effectiveness of the whole system.

In such situations, particularly when the task is to define the characteristics of a major new weapon system, client interest can lead to intense management review, perhaps so intense as to dominate the technical work being conducted to define the system. Multiple layers of review which contribute little to the quality of the analysis require large amounts of time in preparing, presenting, and defending briefings. Conversely, and much more importantly, intelligent client attention can be very productive in helping the analyst develop a model that focuses clearly and narrowly on the real decision issues and in developing the very essential high-level user confidence in the quality of the analysis.

An excellent (but regrettably unusual) example of this occurred during the mid-1960s Air Force studies of the Advanced Manned Strategic Aircraft. At a point in the work, Dr. Harold Brown, then Secretary of the Air Force, began working directly with the study group on a daily basis. Portions of the analysis would be delivered to his office in the early evening and then picked up in the morning with his comments, questions, and suggestions. He well-understood the analysis that resulted after a period of several weeks of such interaction and found it credible. When it came time for him to carry the details to higher levels, the analysis was very sharply focused on both the parameter values and the issues that he knew to be important.

The scale of the effectiveness model is critical. Gross comparisons must be made early on a flexible, intermediate-scale ("medial") model that will test many parameters. The type of model used at this stage of analysis will be rather different from a model used at a later stage to select detailed weapon system characteristics within a particular weapon system class, and neither model is suitable for the other purpose.

This is frustrating to a client who wants to keep the door open to all alternatives. Would an aircraft be better than a tank? Would land-based maritime ASW patrol aircraft or lighter-than-air craft be better than new carrier-based ASW aircraft? Once again, the best results occur when the analyst is able to work closely (and often continuously) with the client to be sure the client understands and accepts the scale of the analysis and the options that are offered and foreclosed by a particular approach to modeling his problem. Very little is more frustrating to the analyst than to complete an analysis he is proud of only to be told that he has "worked the wrong problem."

In the past, a major problem in weapon procurement analysis has been the overestimation of new-system performance. The advocate of a new system is stimulated by both the estimates he sees of possible future enemy capabilities (often based on long extrapolations of incomplete bits of intelligence information) and his natural desire to be responsible for a system that is much better (more capable and/or cost-effective) than current systems. A few years ago, this kind of situation led the President's National Security Advisor to write the Secretary of Defense to ask why US atmospheric offensive systems always penetrated enemy defenses but US atmospheric defenses always defeated the enemy offensive systems.

The problem, of course, turned out to be more one of analysis and of advocacy than of models, but having surfaced the problem, a brief mention of some cures seems in order: (1) long experience with weapon procurement analysis; (2) the analyst's willingness to scale back new weapon effectiveness estimates using operationally realistic degradations; and (3) maintenance of a "devil's advocate" relationship through an outside skeptic.

In brief summary, as described in the Overview, a fundamental purpose of modeling is *communication*. Most importantly, in the weapon selection process that communication becomes *illumination* -- across a broad spectrum of scenarios, tactics, technology, costs/budgets, effectiveness parameters, and so on. The model provides a means for the analyst to describe to the decision maker the boundaries of the problem he is interested in and the trade-offs required for his decisions relative to the weapon or weapon system.

REFERENCES

1. *Robert F. Coulam, Illusions of Choice: The F-111 and the Problem of Weapons Acquisition Reform, Princeton, 1977.*

2. *John M. Danskin, The Theory of Max-Min and Its Application to Weapons Allocation Problems, Springer-Verlag, New York, 1967.*

3. *Gene H. Fisher, Cost Considerations in Systems Analysis, American Elsevier, New York, 1975.*

4. Francis P. Hoeber, Military Applications of Modeling and Selected Case Studies, Gordon and Breach, New York, 1981.

5. E.J. Quade and W.I. Boucher (eds.), Systems Analysis and Policy Planning: Applications in Defense, The Rand Corporation, R-439-PR (abridged), June 1968.

6. The Major Fleet Escort Study, Volumes I, II, and III, Office of the CNO, Navy Dept., 1967.

7. Models, Data, and War: A Critique of the Foundation for Defense Analyses, Comptroller General, PAD-80-21, 1980.

JOHN A. ENGLUND

Mr. Englund has been an analyst since 1956, working mainly on strategic systems in research and development. He has been employed as an operations research analyst by the Strategic Air Command (with targeting, nuclear weapon effects, and nuclear weapon development responsibilities) and by the US Arms Control and Disarmament Agency as a systems analyst with responsibilities for analyzing military systems in an arms control context. He joined the staff of ANSER in 1963 as a mathematician. He has been (successively) Manager of ANSER's Strategic work (1964-76); Executive Vice-President (1976-81) responsible for ANSER's entire technical program; and President (1981-present). He holds a Masters Degree in Mathematics from the Massachusetts Institute of Technology and is a past President of MORS.

ABOUT CHAPTER 9

Drawing on his long experience with Washington decision making, Thomas A. Brown reviews the role modeling has played in force sizing: that is, questions of force level and force mix. During the 1960s it was thought by some that military force sizing could be carried out by a three-step procedure:

- Establish policy goals.

- Determine military capabilities to satisfy the goals.

- Calculate least-cost forces which provide the required capabilities.

Modeling plays an obvious role in the last two steps. However, the complexity of warfare and the wide variety of scenarios which can arise go far to undermine this orderly logic. In addition, political and economic circumstances intrude, in which the adequacy of military forces competes with other espoused needs. Nevertheless, as Brown goes on to show, modeling continues to play an indispensable role in formulating and augmenting the policy maker's appreciation of present and future force capabilities to achieve his policy objectives. His paper outlines characteristics of models to play this more modest, yet still critical, role effectively.

The Role of Modeling in Force Sizing

by

Thomas A. Brown

WHAT IS FORCE SIZING?

In this chapter we will discuss the role of modeling in determining overall force size. Although we will make some brief remarks about the related question of how modeling can help those interested in allocating a fixed budget among weapon systems or in choosing the best weapon system for a given job, the primary thrust is how modeling can help answer questions such as:

- How big should our strategic forces be?

- How much should we spend on our military forces as a whole?

- How large should our Navy be?

First we shall consider the naive view which became popular during the 1960s -- that force sizing can be done rather precisely by applying the appropriate models. Then we shall compare this view with our current understanding and discuss the role which modeling actually plays in force sizing at the present time.

THE NAIVE VERSION

During the 1960s the view was put forward by the Secretary of Defense, Robert McNamara, that force size should be determined as the end product of a three-stage process as follows:

- Determine policy goals.

- Determine the military capabilities required to support the policy goals.

• Calculate those forces which would provide the required capabilities at a minimum cost.

The usual illustration of how this process worked in practice was how we sized the strategic nuclear forces. In this paradigm case, it was determined as a matter of policy that we required forces which would be capable of destroying the Soviet Union as a modern functioning society even after our own forces had received a surprise attack. This was interpreted as meaning that our forces should be capable of destroying 20-25% of the Soviet population and 50% of Soviet industry. In fact, the strategic forces we procured during the 1960s were substantially in excess of this requirement. The submarines alone or the ICBMs alone were capable of destroying 20-25% of the Russian population and 50% of the Russian industry in a retaliatory attack. Furthermore, the standards for assured destruction varied slightly from year to year. Therefore, it is clear that the public version of how force sizing decisions were made did not correspond to what was actually going on.

THE NAIVE VERSION DOES NOT FIT REALITY

It was always admitted by even the most forceful advocates of the view that force size could be calculated as a function of required military capability that although this process might work for strategic forces, it did not work very well for general-purpose forces. For example, consider the following:

> The situation is entirely different for general-purpose forces. Conventional wars generally last a long time. There are literally millions of moves or decision points, in contrast to the one-move strategic war. Even with high-speed computers, it is impossible to construct a model of a conventional war which realistically reflects all the possible alternatives available to each side. Then, too, one cannot predict with confidence the outcome of even small-unit engagements as a function of the opposing forces and arms. The reason is that the results of the engagement often depend critically on factors that are hard to measure or predict, such as tactics, surprise, terrain, training, leadership, and morale. The problem of estimating the outcome of a war in a sector of a theater is even

> more complicated, since that outcome depends on the outcome of many small-unit engagements plus the overall tactics and strategy used by each side. Furthermore, conventional wars in the nuclear age are fought for limited objectives and are as characteristically political as military. In some cases, such as Vietnam, it is not possible to determine who is winning by calculating the number of casualties suffered by each side, or the amount of territory gained or held. Thus, realistically, it is extremely difficult to calculate how much better the United States might do if it placed another infantry division or tank division or wing of tactical aircraft in a particular theater.†

Today it is recognized that the strategic forces themselves cannot be sized in this way. Harold Brown, in his speech at the Naval War College on August 20, 1980, recognized that the requirements for the strategic forces were far more complex than the simple "one-move" exchange postulated by Enthoven and Smith:

> ...deterrence must restrain a far wider range of threats than just massive attacks on U.S. cities.
>
> In our analysis and planning, we are necessarily giving greater attention to how nuclear war would actually be fought by both sides if deterrence fails. There is no contradiction between this focus on how a war would be fought and what its results would be, and our purpose of insuring continued peace through mutual deterrence. Indeed, this focus helps us achieve deterrence and peace by ensuring that our ability to retaliate is fully credible.
>
> By definition, successful deterrence means, among other things, shaping *Soviet* views of what a war would mean -- of what risks and losses aggression would entail. We must have forces, contingency plans, and command and control capabilities that will convince the Soviet leadership

---

†*Alain C. Enthoven and K. Wayne Smith,* How Much Is Enough? *(1971), pages 211-212.*

> that no war and no course of aggression by them that led to use of nuclear weapons -- on any scale of attack and at any stage of conflict -- could lead to victory, however they may define victory. Firmly convincing them of that fundamental truth is the surest restraint against their being tempted to aggression.

This makes it clear that the sizing and selection of nuclear forces involves many of the same problems (multiple scenarios, the possibility of long conflicts, tactics, surprise, leadership, and so on) which plague the sizing and selection of conventional forces. But if the simple "determine requirements, then minimize cost" model of force sizing does not work, how are forces sized in practice?

HOW IS THE DEFENSE BUDGET ACTUALLY SIZED?

During most of the Carter years, the United States aspired to 3% real growth in defense spending each year. This was not achieved because of faulty predictions of the inflation rate. In his final budget, President Carter urged a 5% real annual growth rate; now President Reagan is urging a 7% annual growth rate. The fact that these real growth rates are round numbers and are projected to be constant indicates that the real decision, which is made at the top, is not to buy certain forces and then sum the cost of these forces to see how much you have to spend, but rather a decision on how much the country can afford to spend. It is then left up to the experts in the Department of Defense to work out how the total budget should be allocated amongst the competing forces. But how is this real growth rate selected? What makes the President decide that 0% or 3% or 5% or 7% is required? It appears that there are a number of factors which enter into this. These factors can be divided into general categories. On the one hand are domestic factors such as the probable tax revenue, the state of the economy, the projected budget deficits, and other considerations which indicate how much we can afford or are politically willing to spend. On the other hand we have the military factors, the world balance of forces. How is the military balance measured? There are three basic ways in which the military balance between ourselves and the Soviet Union is assessed:

- Side-by-side comparisons (including Allies).
- Budgetary comparisons.
- Assessments of probable war outcomes.

The first kind of comparison, side-by-side comparisons of force structure, is found in newspapers and magazines all the time. Sometimes these side-by-side comparisons are quantitative and sometimes qualitative. One difficulty of side-by-side comparisons is that the Soviets are ahead in some categories and we are ahead in others. You can make the Soviets look stronger by selecting categories in which they are ahead, or you can make our side look stronger by selecting categories in which we are ahead. Another problem with this kind of comparison is that it usually ignores initiative and all other dynamic considerations, and these considerations have a very great impact on what would happen in an actual test of strength. It is difficult on the basis of side-by-side comparisons of individual weapon systems or individual aspects of the forces on the two sides to reach a conclusion as to our overall relative degree of strength.

Budgetary comparisons are one way of attempting to gain an insight into the total strength of the two sides. It is difficult to get a meaningful comparison between U.S. and Soviet defense budgets because much of the defense spending on the Soviet side is secret, or hidden in budget categories other than the Ministry of Defense. Furthermore, it is hard to compare rubles to dollars since there is not an international free market in "defense" from which the two sides make common purchases. Finally, there is the fact that "inputs" do not necessarily correspond to "outputs," and it is conceivable that one side could pour massive resources into defense and realize very little return, while the other side expended more modest resources but managed to maintain substantially more effective forces. The CIA† has attacked these three problems by means of the following technique: they consider the Soviet forces actually deployed and analyze what

†*National Foreign Assessment Center, Soviet and U.S. Defense Activities, 1971-80: A Dollar Cost Comparison, 1981.*

it would cost us to deploy similar forces in this country. It is well-known that this tends to exaggerate the size of Soviet forces somewhat because they select their forces on the basis of their relative cost within their own economy, and thus to price these forces within the context of our economy will exaggerate their value somewhat. For example, the Soviets have a very large army. Much of the army consists of conscripts who are paid very little. It would cost us, in our all-volunteer force, considerably more than it costs the Russians to maintain a force of that size. In dollar terms (that is, measuring Soviet forces in terms of what equivalent forces would cost in this country), the Soviets have been spending 50% more than the United States in recent years. In ruble terms (that is, measuring U.S. forces in terms of what equivalent forces would cost in the Soviet Union), the Russians have been outspending the U.S. by about 30%. Just as the former measure exaggerates Soviet forces, so does the latter measure exaggerate U.S. forces. Perhaps the greatest value of budgetary comparisons, however, is in the measurement of long-term trends. The fact that the Soviets have been steadily increasing their forces in real terms by 3% a year since about 1968, while our own forces have been going down during much of that period and rising by less than 3% during the balance of the period, certainly gives us some cause for concern. Even so, budgetary comparisons are not a very satisfactory measure of relative military power. As the CIA document referred to above states:

> Dollar valuations, however, still measure input rather than output and should not be used as a direct measure of the relative effectiveness of U.S. and Soviet forces. Assessments of capability must take into account strategic doctrine and battle scenarios; the tactical proficiency, readiness, and morale of forces; the numbers and effectiveness of weapons; logistic factors; and a host of other considerations. Thus, while dollar valuations can portray changes in the military emphasis of a nation's forces over time, they are not sufficient alone to compare the capabilities of U.S. and Soviet forces.†

†*Soviet and U.S. Defense Activities, 1971-80*, page 11.

This brings us to our third approach, assessment of probable war outcomes. This is where modeling comes into play. Modeling of actual combat is very difficult for all the reasons cited in the Enthoven and Smith quotation above. However, combat modeling is probably the most important of the three ways of assessing the adequacy of our military forces because it is most relevant to what will happen if the forces are ever actually put to use. It has been said that we procure military forces for three reasons:

- To deter war.
- To win a war if war occurs.
- To extend our influence in peacetime.

The mission of deterrence and the mission of extending influence depend critically on the perceptions of others as to what would happen if war actually occurs. Combat modeling is the technique by which we gain some insight into what the outcome would be if the forces were actually put to use. Therefore, it is the fundamental way in which the adequacy of our military posture must be assessed. However, combat modeling is difficult and fraught with uncertainties. Wars could occur under a wide variety of scenarios, and therefore the calculation of the required military forces on the basis of combat modeling can never be reduced to a simple formula.

There are three criteria which combat models should meet (in addition to the fundamental criteria of accuracy and realism) if they are to be useful in force sizing:

- Transparency,
- Reproducibility,
- Flexibility.

Transparency is important if the model is to be credible. No rational decision maker is likely to conclude that our forces are adequate (or inadequate) on the basis of some mammoth computer model which he does not understand. The modeler must try to make clear both what happens in combat and also why it happens. The quality of transparency not only enhances the credibility of the model, but also enables the model to be used to suggest improvements which should be made in the force posture in order to correct its deficiencies.

Reproducibility is an essential characteristic of scientific experiments and a desirable characteristic of combat models. It is necessary if results of one model are to be of use in testing, verifying, or improving the results of another. There are, however, some useful models for which it is very difficult to achieve a high degree of reproducibility. Games involving human players will give different results for different teams, even if they start from the same scenario. You can, however, secure a kind of reproducibility even for models of this kind. Careful documentation of each decision made on each side (and equally careful recording of any random numbers drawn in evaluating the consequences of the players' decisions) should enable any given run of a game to be reproduced in full detail. Such thorough documentation is a burden, however, and it is not worthwhile in some applications (such as when a game is run for training purposes or strictly for the enlightenment of the players themselves).

Flexibility is required so that the model may be used to examine as wide a variety of scenarios and tactics as possible. One of the dangers of large models is that they may limit players to a stereotyped collection of possible actions. In this way, very serious vulnerabilities may be totally overlooked. For example, many models of strategic war have made the implicit assumption that a "bolt from the blue" attack is the "worst case," and if our forces can do grave damage to the Soviet Union after such an attack they are adequate. In fact, the results might be even worse if a nuclear war would start after an extended period of conventional conflict in which our communications and warning were seriously degraded (either through sabotage or direct attack) and our assets deployed afloat or in Europe were to suffer significant attrition even before the first nuclear weapon was used.

Models which combine reasonable realism with transparency, reproducibility, and flexibility not only can be used to assess the adequacy of our forces with respect to those of potential enemies, but they also give indications of why our forces fall short and just where they most need strengthening. Therefore, they serve the needs of force design as well as force sizing.

The distinction between "force sizing" and "force design" often breaks down in actual application of models. For example, in 1962 a game called Strategy And Force

Evaluation (SAFE) was created at The Rand Corporation for use in studying the impact of alternative central war strategies on force posture. In six plays of the game, players took various policy statements and budgets and built force postures. In every case but one, the players expended the entire budget they were allocated. There seemed to be little linkage between policy statements and postures (that is, weapon systems preferred under one policy statement were likely to be preferred under any policy statement). Each SAFE "play" covered the period 1962-72, with a war at the end to test the postures the two sides had created. A retrospective analysis in 1975 revealed that the postures created in the games played in 1962 were surprisingly close in structure to the postures created by the United States and the Soviet Union over the actual period from 1962 to 1972. Those discrepancies which did exist could be traced to specific shortcomings in the menu of strategic systems on which the game was based. Therefore, the SAFE game provided little guidance on force sizing (since the players tended to build as big a posture as they were permitted to build) but provided a reasonable representation of the force design problem.

During the 1960s some force planning models were constructed which would calculate how much support was required, given a desired force posture, to maintain the historic force/support relationships. Given the political dynamic, which seems to always lead to cuts in support rather than in force posture, the actual support is almost always less than the historic relationships. This is all that models such as the Navy Resources Model told us. It would be better if a model would go forward and give indications of what the shortfall in support would mean in actual combat. This is the information required if we are to make intelligent trade-offs between force size and support.

A general problem which afflicts the use of combat models is the problem of packaging the results. A skeptical decision maker is not going to accept the results of someone else's war game at face value. He knows the results will depend on the scenario, on the tactics used on the two sides, and on numerous submodels whose verisimilitude may or may not have been carefully audited. Nevertheless, there seems to be general agreement that war gaming generates insight among those who are directly involved in it. Therefore, there appear to be two ways out of this problem:

- The war gamers may find ways to present their insights independently of the games in which they were originally conceived.

- War games may be designed which are transparent and so easy to play that fairly high-level people may be directly involved in playing them.

Both of these approaches require considerable creativity on the part of analysts. But who ever said that systems analysis was a pedestrian occupation?

In summary, combat modeling attempts to tell us how our forces would do in certain scenarios. Combat modeling is also an aid in calculating how much we would have to spend to achieve certain results. On the basis of such calculations as to how we would do if war occurred and how much it would cost to improve the likely outcome, repeated for a wide variety of scenarios, it is possible to form a general view of whether the increase in deterrence, security, and influence gained by an increase in defense posture would be worth the cost. The decision to increase or decrease the real defense budget at a certain pace is undoubtedly based, in part, on general views formulated in this way. Therefore, modeling of force interactions plays an important indirect role in force sizing, even if it does not play the direct simplistic role presented in some of the writings of the 1960s.

REFERENCES

1. *Alain C. Enthoven and K. Wayne Smith, How Much Is Enough?, Harper and Row, New York, 1971.*

2. *Soviet and U.S. Defense Activities, 1971-80: A Dollar Cost Comparison, National Foreign Assessment Center, Washington, January 1981.*

DR. THOMAS A. BROWN

Dr. Brown was Deputy Assistant Secretary of Defense (Strategic Programs) for the period 1977-81. He attended Oxford University on a Rhodes Scholarship and holds a Ph.D. in Mathematics from Harvard University. He worked for many years at The Rand Corporation and at Science Applications Incorporated, and is now Director of Strategic Studies, Washington, for Rand. He has published numerous articles on defense, operations research, and mathematics.

ABOUT CHAPTER 10

If national policy analysis deals with defense-related issues that transcend decisions based purely on military considerations, then modeling in the broader sense has been used to help study national policy issues for many years. But a significant milestone was passed in the 1960s when thoroughgoing analytic and simulation models were constructed explicitly to aid in the analysis of arms control negotiations, in which defense, foreign policy, and perhaps other points of view all became mingled. Alfred Lieberman illustrates some of the unique properties and problems that intrude when these exceedingly complex and far-reaching problems are considered. As a second example of meta-military modeling, he also discusses U.S. efforts to estimate Soviet defense expenditures.

The parallels and contrasts found in his two examples adequately indicate, we believe, the difficulties and the opportunities in national policy analysis. The thoughtful reader will grasp more readily by reading Lieberman's dissertation why contentiousness has abounded in both areas, and can justifiably reach the conclusion that any analysis attempting to shed light on other issues having so diverse a set of partisan viewpoints will probably be similarly controversial. Still, here as elsewhere, models have helped to pinpoint and structure debates. Lieberman illustrates for our readers an important arena of modeling about which many may heretofore have thought very little.

Models in National Policy Analysis
by
Alfred Lieberman

National policy analyses cover a spectrum of issues which include strategic nuclear policies such as mutual assured destruction, damage limitation, limited nuclear war, flexible response, and protracted nuclear war. Early uses of models in national policy analyses were in force costing and in the studies of civil defense which led to the civil defense debate of the 1960s. These subjects have been covered in the chapters on nuclear warfare and force sizing and will, therefore, not be addressed in this chapter.

Other foreign policy issues such as arms sales to foreign governments, assessments of world oil supply, the grain embargo, and the freezing of financial assets have also been modeled. In addition, some work has been done on the modeling of negotiations and the use of models as a negotiation tool.† However, the work on these policy issues has been relatively limited and should be a fruitful field of endeavor for the 1980s.††

This chapter will concentrate on two applications: models used within the Federal Government in support of arms control analyses and those used in support of the

---

†*See, for example, the work done by Nigel Howard in* *Paradoxes of Rationality: Theory of Metagames and Political Behavior*, *MIT Press, 1971, and Thomas L. Saaty's* *Mathematical Models of Arms Control and Disarmament*, *Wiley, 1968.*

††*Models are also used in state and local governments. See, for example,* *Models in the Policy Process: Past, Present, and Future*, *by Warren Walker, The Rand Corporation, September 1981.*

cost estimation of Soviet military expenditures. While there are, of course, many other models used outside the Government in support of these two policy areas, a discussion of models used within the Government should be sufficient to illustrate some of the key points in the development and use of models in policy analyses.

I. MODELS IN ARMS CONTROL ANALYSES

Introduction

The modeling of national policy issues was not widely done outside the DoD until the 1960s, when the strategic arms control negotiations became a central feature of our foreign policy. Analyses using models were performed by the U.S. Arms Control and Disarmament Agency, the Department of Defense, and numerous contractors in the development of the U.S. position for the ABM Treaty and for the SALT I Interim Agreement. Some of the key issues in this period for which models were intensively used dealt with the development and deployment of MIRVs and with the ABM debate. In this latter issue, both sides in the debate made use of results obtained from model calculations; and interestingly enough, both advocates and adversaries of the ABM used valid models but came to opposite conclusions. The differences were due to the choice of assumptions made, for which the range of uncertainty was large. Assumptions and input data are often the key issues in the use of models, and this has strongly influenced the way models are used in analyses.†

In the 1970s, the use of models to support the SALT II arms control negotiations became even more pronounced. Agreement on how to resolve many issues was possible only after analysis with a model had revealed the interactions of the variables. Examples of such analyses are: the best way to measure the strategic balance, the potential vulnerability of ICBMs, and the possible impact of civil defense on deterrence. During this period, policy analysis models were also developed for two other issues.

---

†*An understanding of the impact of changes in input data can be seen on page 109 of "Choosing a Strategy for World War III," by Thomas Powers, The Atlantic Monthly, November 1982.*

These were the Soviet view of analytical procedures† and the economic and cost models developed by the intelligence community and others to estimate the cost of Soviet military expenditures.

In the 1980s, models continue to be an important factor in the arms control negotiations. In fact, one of the most crucial issues in the development of the U.S. START position was the selection of the units of limitation to be used in the agreements. Since some of the units of limitation were also selected by the analysts as measures of effectiveness, care had to be exercised to ensure that the results were not a product of circular reasoning.

The Selection of Models

In the 1960s, analysts at the U.S. Arms Control and Disarmament Agency began searching for models which could be used in support of the development of negotiating positions for the forthcoming strategic arms limitation negotiations. In order to understand the complexity of this task, one needs only to look at the latest JCS catalog of models††, which is over 800 pages long and contains descriptions of almost 400 models. The choice of models in the 1960s was similarly extensive.

The first question addressed in the search for a model dealt with the level of detail or fidelity required in models for arms control policy decisions. It was clear from the beginning that a detailed model such as QUICK, which is used by JCS/SAGA for the RISOP/SIOP evaluations, was neither required nor feasible for models supporting arms control negotiations. The great number of details which are of importance to the JCS for developing the specific utilization of individual weapon systems, tactics, and strategies were not needed by ACDA. Furthermore, the manpower required to set up the input data for such a model and the 6-hour average computer run time for each case were far beyond the resources that ACDA could devote to analysis. On the other hand, models such as the Arsenal Exchange Model (AEM) and Code 50, while valuable for overall force sizing studies in DoD, had a number of

---

†Particularly noteworthy was the work performed by Allan Rehm at the CIA, and J. Battilega and J. Shannon at Science Applications, Inc.

††Catalog of Wargaming and Military Simulation Models, SAGA/JCS, May 1982.

shortcomings. The Arsenal Exchange Model, which was intensively used by OSD, was particularly useful in designing near-optimum forces for specific strategic objectives. The results, however, were too complex for the decision maker, and since AEM did not include geography, it did not permit the assessment of collateral damage.† Code 50 had many outstanding original features, such as the use of Lagrange multipliers for weapon allocation and the capability to recall previous runs. It too, however, did not explicitly include geography and therefore did not permit assessment of collateral damage; moreover, it was only a one-sided attack model. Even before the start of the actual negotiations, it was clear that two-sided attacks had to be considered. Furthermore, the interest in the effects of nuclear war and in civil defense made it imperative that collateral damage and fallout effects be included.

A number of other factors had to be considered in selecting the models used in arms control analyses. First of all, arms control agreements take a long time to negotiate; and once negotiated, they may apply for long periods of time. For this reason, the analyses do not deal with today's force but with forces of the future and sometimes of the far future. Only the most general characteristics of these future forces can be predicted. The analysis must, therefore, be designed to study a broad range of alternative assumptions -- not only about such things as missile accuracy and silo hardness, but also about new weapon systems which may be designed to exploit ambiguous wording or incomplete constraints in an arms control agreement. Treaty verification and breakout add to the uncertainties which must be modeled. That is, the analysis must estimate the changes in the results which would occur if certain provisions of an arms control agreement were violated and not detected. Similarly, the potential for rapid breakout from an agreement must be examined. Breakout is a term used in arms control which refers to the potential of either side to suddenly abrogate an arms control treaty and rapidly break out of the treaty constraints.

---

†*A new AEM model is currently being developed by SAI. It will include detailed geographical modeling and collateral damage calculations.*

On the basis of these requirements, ACDA selected the SIRNEM† model -- parts of which were later incorporated into a more efficient model called RPM. SIRNEM/RPM is really a group of strategic models arranged so that each module can be used as needed for a particular study. This action-by-request feature allows a wide variety of scenarios to be studied. In addition, ACDA also used the Arsenal Exchange Model for a number of years to examine optimum force structures under various arms control constraints.

SIRNEM/RPM simulates strategic warfare on an intermediate level of detail and provides great flexibility in both the level of detail used and the analyses performed. In a typical study, each side might make war plans based on prewar information (which can be different for the two sides) and on estimates of wartime conditions (such as the amount of damage which forces might sustain before the weapons are launched). A retaliatory attack can be launched using either of these two prewar plans or using the forces actually calculated as surviving the first attack. The latter implicitly credits the retaliation with a retargeting capability.

SIRNEM/RPM modules take location and range constraints into account so that collateral damage and target accessibility can be assessed. The flexibility of these models permits the assessment of either blast damage alone or of both blast and fallout damage. Bomber penetration can be incorporated by just a few input parameters or by a module which simulates bomber penetration in great detail. These flexible features have proved very important since they permit studies of fallout and detailed bomber penetration to be included when required, while also permitting large savings in time and money when they are not required.

Acceptance of Models

Another important issue which arose in the analyses of SALT agreements was the fact that some of the modeling work was done for decision makers in the State Department and the National Security Council who were much less comfortable with quantitative analyses than their counterparts in the Department of Defense. Furthermore, the extreme political sensitivity of arms control made decision makers suspicious of models and analytical techniques

---

†SIRNEM and RPM were developed by the Academy for Interscience Methodology (AIM).

which they could not fully understand. These suspicions came to a peak when questions were raised as to whether or not a particular model was only applicable when the underlying assumption of the strategic doctrine was that of mutual assured destruction.

This situation was a clear demonstration of the importance of transparency in models. Models at even an intermediate level of detail contain enough structure so that they were fairly difficult to explain to high-level decision makers who had little time and limited experience in quantitative analyses. It became clear that the analysts themselves had to understand the models quite deeply in order to properly explain them. Careful explanation did, in part, alleviate some of the decision makers' suspicions of the models. Another factor that helped was that the SIRNEM and RPM models were also being used by the Navy and by JSTPS. The fact that three different organizations doing substantially different work were all using these models tended to enhance each organization's confidence in the models. This multiple use of the models provided a cross-check of the basic logic and structure of the models.

Perhaps the most convincing evidence for the acceptance of the use of these models came from the fact that in the arms control arena all issues are resolved on an interagency basis. The problems which were studied with the ACDA models were also run separately by OSD and JCS, thereby providing independent results using different models. Generally, the results were found to be in close agreement; when they did not agree, the differences were almost always due to differences in assumptions or in input data. Making such parallel studies has great benefits -- not so much for the few occasions when a modeling or analytical error is found, but for the much more frequent occasions when real light is shed on the different underlying view which each organization had on a specific problem. The replication and comparison of analyses by different organizations in the bureaucracy provides perhaps the best way to clarify the disagreements between the various bureaucratic entities. Almost without exception, recent disagreements can be traced to differences in assumptions, the amount of risk different proponents would accept, or the input data -- not to differences in the models. The resolution of such data assumption issues by bureaucratic political processes has become an important and valuable part of modeling.

Model Requirements Evolve

These models were extensively used by ACDA to study a wide variety of problems. In the SALT I preparations, the models were used in the evaluation of the impact of MIRV and the potential effect of the ABM. In SALT II preparations, the models were used in the evaluation of options and in the interagency civil defense study which led to PRM-32.†

While the SIRNEM and RPM models were extremely useful and continue to be of great importance, it became clear that more aggregated models were also needed. The reason for this is that the number of cases which had to be investigated was so large that even intermediate-level models such as SIRNEM/RPM were too detailed.

The proliferation of cases to be studied was a result of a number of factors:

- Arms Control Options. In the SALT II and in the START preparations, the number of options to be investigated was extremely large. Frequently, new options were generated at a more rapid rate than old options could be evaluated.

- Uncertainty. Negotiations like SALT II and START take years to negotiate and once negotiated may stay in force for many years. The analyses of the negotiations, therefore, do not deal with today's forces but with those of the 1990s. There is a lot of uncertainty in the characteristics of the U.S. forces of the 1990s. When one considers that the analyses also require that the characteristics of the Soviet forces and the target data bases for that time frame be estimated, then one can appreciate the meaning of uncertainty. In order to investigate the impact of these uncertainties, analyses must be made using a range of values for each of the important parameters. The cross product of all these parameters can yield a large number of cases. If each of the arms control options generated above were to be investigated over the whole range of uncertainties, the number of cases to be studied would rapidly grow out of bounds.

---

†*Presidential Review Memorandum No. 32 on Civil Defense.*

• Verification. Treaty verification is a very important issue in arms control. It is obviously not possible to verify every aspect of an arms control agreement with 100% certainty. It is essential, however, to know what the potential violations might be and what impact such violations might have on the nuclear weapons balance. In order to determine this impact, it is necessary to estimate all the potential violations and to analyze each case, thus greatly multiplying the number of cases to be studied.

• Breakout. The most important aspect of breakout is that we must maintain the capability to take adequate counteraction to an abrogation of an arms control agreement by the Soviet Union before the breakout has a major impact on the strategic balance. What is required, therefore, is to determine the breakout potential of the Soviet Union and to estimate the time it would take us to react. The strategic balance during the intervening time period must then be analyzed to ensure that major destabilization cannot occur. This further increases the number of cases to be studied.

Measures of Effectiveness

Along with the proliferation of different cases which need to be examined, there has also been a proliferation of measures of effectiveness, each of which at least some of the participants in the arms control arena wanted to include. The measure-of-effectiveness problem is not a new one; analysts have been working on this issue since the early days of operations research. In arms control issues, however, this problem took on special meaning for two basic reasons:

1. Some measures of effectiveness showed that the strategic balance favored the U.S.S.R., and some showed the contrary.

2. In some cases, the units of effectiveness selected themselves became a quantity to be limited under the proposed arms control agreement. This factor tends to indicate which features of the forces the policy makers consider most important. Differences in opinion during the negotiations on the selection of the units of limitation tend to indicate the different views the sides have on the most pertinent features of the forces.

In START, the selection of the measures of effectiveness and the units of limitation actually became a major issue. In SALT I and II, the basic unit of limitation was the number of launchers. Analyses performed in the preparation of SALT I were generally quite sophisticated and examined all the various measures of effectiveness in great detail. Generally speaking, people were not dissatisfied with the strategic balance that resulted from these analyses during the negotiation period. When the agreement was signed, the basic unit of limitation was the number of launchers; it was fixed at the number that both sides had at that time. During the ratification of SALT I, however, attention focused on this unit of limitation and, since the Soviet Union had more launchers than the U.S., objections were raised to the equity of the treaty. In SALT II, the basic unit of limitation was again the number of launchers. Care was taken this time to set the limit at an equal number of launchers for both sides, but objections were raised to the fact that imbalances existed in other measures of the strategic forces.

In START, it was felt that the number of launchers was an inadequate unit. Limitations are also required on the number of warheads and on their size. Warheads and throw weight, which were peripheral in SALT, have become the central units in START. Furthermore, not all warheads are alike. Missile warheads, SLBMs and ICBMs, because of their rapid delivery potential and their increased accuracy, are considered to be more destabilizing than weapons carried on bombers. The vulnerability of ICBMs is also viewed as contributing to instability, as is the deployment of MIRVed missiles. The limitation of ballistic missile warheads, particularly ICBM warheads, therefore received priority attention in START preparations.

The consideration of these factors affected both the treaty's units of limitation, where the number of ballistic missile warheads became a limit, and the measures of effectiveness, where Prompt Hard Target Kill (PHTK) became more important than Hard Target Kill (HTK). PHTK is a better measure of surprise attack potential than is HTK.†

---

†Hard Target Kill is the summation over all warheads in the force of the hard target kill probability (for example, 2,500 psi hardness), taking into account CEP and reliability. Prompt Hard Target Kill is like HTK except summed over ballistic missile RVs only.

These factors required that the results of analyses be presented using a large number of measures of effectiveness and that the model be able to rapidly generate all these measures. Here we can only suggest some of the complications which arise when explaining results which measure a phenomenon in a dozen different ways. First of all, it is important that the measure of effectiveness does indeed measure a facet of the issue which is relevant to the problem under consideration. Second, the measures should be relatively independent of each other and, where they are not, the interdependence should be made clear in the analysis. This is not always easy to do and at times requires some judgment. For example, hard target kill capabilities against targets of 2,000, 3,000, and 4,000 psi hardness tend to measure the same phenomena and should not be evaluated as three separate measures of force capability. On the other hand, to measure the same force against 50 psi targets presents a somewhat different aspect of the force and could be considered as a different measurement. Another important consideration is a clear understanding of the meaning of a measure. For example, the computation of hard target kill against 2,500 psi targets could show that a force has the capability to destroy 7,000 such targets. This measure, however, implicitly assumes that there are a sufficient number of targets of this hardness available so that one weapon can be used on each. If, however, the opponent possesses only 1,000 targets of this hardness, then the meaning of a hard target kill capability of 7,000 becomes ambiguous at best. In actual use, several weapons would be allocated to each of these targets and, in any case, not more than 1,000 hard targets could be killed.

An Aggregated Model

In order to handle the large number of cases and the many measures of effectiveness, an aggregated model called FOREV† was developed. This model was specifically designed to compare arms control options. The model requires that the force structures and weapon characteristics of the U.S. and U.S.S.R. be input. The model will first compute a series of static indices describing the pre-attack forces. Following this, the model will calculate counterforce exchanges. That is, calculations

†The FOREV model was developed by A.J. Hartzler of the U.S. Arms Control and Disarmament Agency.

will be made of the damage done and forces left after a counterforce attack by one side and a counterforce retaliation by the other.

The results from the counterforce exchange are then summarized and presented in "fish" curves. The "fish" curves are a graphic representation of a nuclear counterforce exchange; they show the effect of each side striking first against the other side's counterforce targets and the other side retaliating against the attacker's counterforce targets. The model then recomputes all the static indices using the forces left after the counterforce exchange.

An example of a "fish" curve is shown in Figure 10. In the chart the number of warheads in U.S. forces are plotted along the Y-axis, while the number of warheads in Soviet forces are plotted along the X-axis. Therefore, whenever the graph is above the diagonal central line, the balance is to the U.S. advantage; and whenever it is below this central line, the balance is to the Soviet advantage. Please note that the balance in this chart is measured in terms of the number of warheads, and that similar charts could be constructed with other indices.

The nuclear exchange calculations start from the dot in the upper right-hand corner of the figure. This is the balance of the forces before the attack. In the figure, it represents approximately 11,000 weapons for the U.S. versus 9,000 for the U.S.S.R. The chart then shows two possible situations. One of these is a Soviet first strike (dash-dot-dash line) against U.S. strategic forces (ICBM, SLBM, and bombers) and a U.S. retaliation (dashed line) against Soviet strategic forces. The other situation depicted in the chart is a U.S. first strike (solid line) against Soviet strategic forces and a Soviet retaliation (dashed line) against U.S. strategic forces.

The forces remaining after this counterforce exchange are the forces that can then be used for attacks on other military targets and on industrial targets.

The separation between the end points of the two dashed lines gives a measure of the advantage each side would gain by a first strike, assuming that an attack was certain. The larger horizontal separation at the end points indicates that the U.S.S.R. would lose about 2,000

Fi

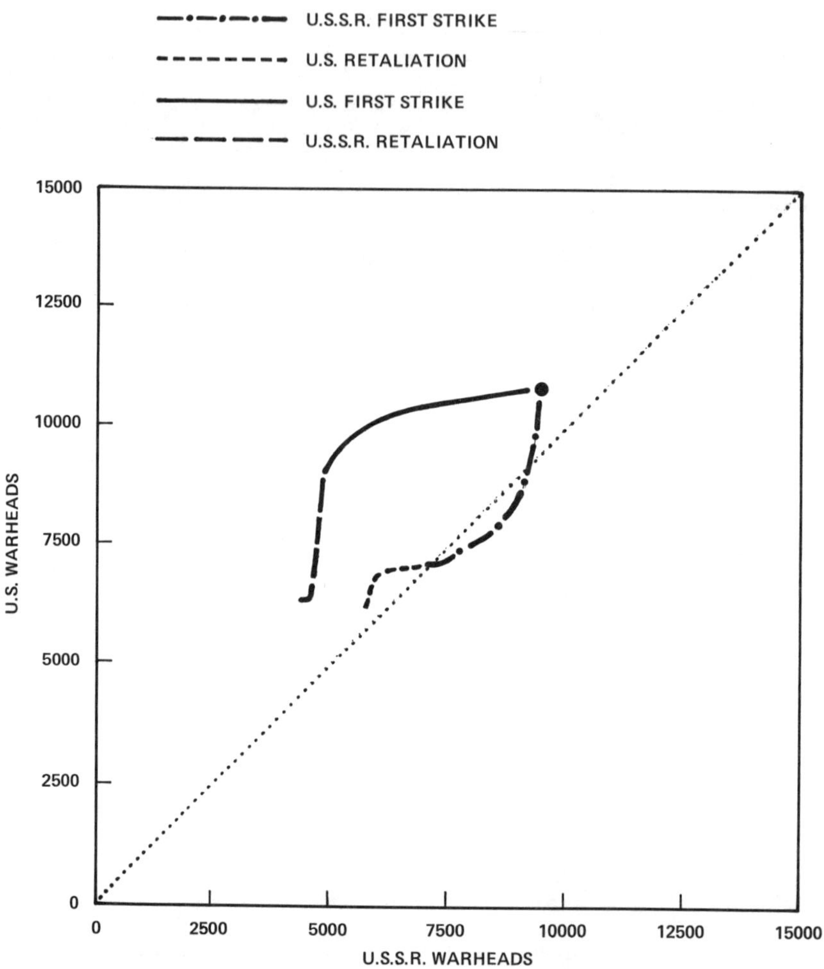

Figure 10. Strategic Counterforce Exchange (Generated Alert)

fewer warheads if it attacked first. In the vertical direction there is little separation, indicating that the remaining U.S. forces are roughly the same regardless of who strikes first.

The analysis is normally concluded by running the countervalue module of FOREV, which simulates attack against the other side's military and economic targets. This countervalue simulation is performed using a series of empirical equations obtained from numerous countervalue attacks performed with the RPM model.

Uncertainty -- A Special Case

A special case in the handling of uncertainty arose in the analysis of Minuteman vulnerability in the late 1970s. This analysis was performed on an interagency basis, with several agencies independently doing the calculations. The input data were supplied by the intelligence community. The models used by the various agencies were roughly similar.

The calculation of ICBM vulnerability requires a number of input parameters. The principal parameters are the hardness of the silo, the yield of the attacking weapons, the accuracy of the weapons, various reliability factors, and the number of attacking weapons which can be effectively used against each silo. The intelligence community always provided a "best estimate" and an upper and lower value to indicate the range of uncertainty for each of these input parameters. The calculations were then performed by setting all the inputs to the best estimates, then all to the upper value, and then all to the lower values. This is the procedure which had been used in the past and which was at that time the accepted method for handling uncertainty. A problem arose, however, in that the band around the resulting number of Minuteman silos which were expected to survive the attack was so wide that essentially the results were of little use. This is because the upper and lower limits calculated by this procedure have an extremely low probability of occurrence.

In order to overcome this difficulty, a different approach to handling these uncertainties was proposed.† It was reasoned that it was most unlikely that all the input parameters would be at either the upper or lower end

†Paul Davis, who was then at PA&E, was the first to suggest the use of this procedure for this application.

of the uncertainty range. It was most likely that some were overestimated while others were underestimated. The intelligence community was then requested to provide a distribution for each of the input parameters.

Models were then built both in ACDA and PA&E which would take a number of discrete intervals with their associated probabilities of occurrence from each of these distributions. All possible combinations of the midpoints of these intervals were then used in the calculation of the number of Minuteman survivors. The probabilities of occurrence of each such combination of the values were also computed. The results were then combined to generate a probability distribution of the number of Minuteman survivors. Confidence intervals around the mean number of survivors could then be obtained from these resulting probability distributions.

The results obtained from this model provided tighter intervals than those that had been obtained in the past. Furthermore, the basis of this model was sounder and the results more useful than those obtained with the use of extreme values. It should be noted that this procedure also contains some hidden assumptions, such as the assumption that all the parameters are independently distributed. Nevertheless, this approach was a positive step forward.

II. MODELS FOR ESTIMATING SOVIET DEFENSE EXPENDITURES

Introduction

The Soviet Union treats defense spending information as a closely guarded state secret. The Soviets report only a single-line entry for defense in their published state budget, and this figure is obviously only a fraction of their actual expenditures. Consequently, analysts both in and out of the U.S. Government have developed their own estimates of the true cost of Soviet military activities. Some of these estimates are reported in rubles, some in dollars, and some in both. All such estimates, however, are necessarily based on analytical constructs which are subject to errors and limitations.

Two principal methods† are used to estimate how much the Soviets spend on defense. One method, known as the Residual Cost Model, infers defense costs from published Soviet economic statistics. This is the approach which most non-governmental researchers have taken. The other method, known as the Direct Costing Model, is the building-block approach and is used only in the intelligence community because of the mass of data needed. Physical components of the Soviet defense effort are identified and enumerated over time, after which direct cost factors are applied to each component. Although any methodology is uncertain, the intelligence community finds the building-block approach more reliable and useful.

Residual Cost Model

The Residual Cost Model†† identifies Soviet defense expenditures from published statistics through the elimination of civilian consumption and exports. The estimates of Soviet defense expenditures also include some "best-guess" estimates (particularly for the RDT&E portion, which is arbitrarily increased by about 30%).

The major weakness of the Residual Cost Model is its dependence upon statistics supplied (and perhaps slanted for propaganda purposes) by the Soviets. Additionally, this model produces only a ruble cost of Soviet defense expenditures and would, therefore, require translation into dollars for the purpose of comparison with U.S. outlays. The ruble estimate alone, however, does permit the tracking of Soviet budget changes and the assessment of the relative importance placed on defense spending by the Soviets over time.

The proponents of this model contend that Soviet statistics, if properly interpreted, are sufficiently accurate for the purpose. The model also has the advantage that it is relatively simple and has proven to be surprisingly accurate when an independent check of Soviet defense expenditures was available.

---

†*Some other models have also been proposed such as Methodology for Estimating and Comparing Soviet Defense Expenditures, by Robert Allan Walker, USC, 1979.*

††*The best of these models is attributed to William T. Lee, currently at DIA and an analyst at the CIA from 1951 to 1964.*

Direct Costing Model

In the Direct Costing Model, the estimation of Soviet defense costs begins with an item count of hardware, installations, and personnel. Costs for a given item are then derived by applying engineering estimates, parametric cost estimating relationships, and analog costs -- all based upon the hypothetical production or employment of the item in the West. These estimates are adjusted ("Sovietized") to account for the different military, resource, and manufacturing conditions known to prevail in the Soviet Union. Individual item costs are then aggregated to form an estimated actual Soviet defense budget commensurate with U.S. budgets. As a final step, the estimate is checked for reasonableness against published Soviet statistics (including the data used in the Residual Cost Model) and interpretive data collected through other means (e.g., covert operations).

Costs are developed for all Soviet defense activities except RDT&E by identifying and listing Soviet forces and their support apparatus. The model contains a description of about 1,100 distinct defense components -- for example, surface ships, ground force divisions, and air regiments -- and the latest estimates of the order of battle, manning, equipment inventories, and new equipment purchases for those components. The descriptions and estimates reflect contributions from analysts throughout the intelligence community. They are the best available data base of qualitative and quantitative information outside of the U.S.S.R.

Appropriate ruble or dollar prices are then applied to these detailed estimates of physical resources. The dollar estimates are developed using prevailing U.S. prices for all funding categories except RDT&E, which is calculated in rubles and then converted to dollars. To develop the estimate of total Soviet defense expenditures expressed in rubles, costs of RDT&E, construction, personnel, a portion of O&M, and of procurement are estimated through the direct application of appropriate ruble prices. The remaining categories of defense spending are calculated in dollars and converted to rubles using carefully constructed ruble-dollar ratios.

Dollar and Ruble Cost Comparisons

To state one country's activities in terms of another country's currency will exaggerate somewhat the size of size of the first country's effort. This phenomenon -- called the index-number problem -- has been the basis of some criticism of the intelligence community's attempts to compare Soviet and U.S. defense costs in terms of dollars. Because of the index-number problem, some critics argue the dollar comparisons of U.S.-Soviet resource commitments to defense should be supplemented with comparisons made in ruble terms.

As a result of differences in resource endowments, the U.S. and Soviets have different relative cost structures in their economies. When the Soviets make their defense budget decisions, they choose their defense mix considering their own costs. If they were instead faced with the U.S. cost structure, they might make different choices -- choices which would produce the same utility for them but at a lower "dollar" cost. Valuing one country's choices, then, in another's currency tends to result in an overly high price tag because it does not consider the potential cost savings that would result if the first country's leaders had been able to make their decisions under the "new" cost structure. Dollar comparisons of U.S. and Soviet defense activities tend to inflate the size of Soviet costs relative to those of the United States. Manpower is relatively more expensive in the United States. Therefore, dollar comparisons somewhat exaggerate the cost of Soviet defense activities, which are more manpower-intensive than those of the United States. Comparisons made in ruble terms have the opposite effect. In the Soviet Union, capital goods are relatively expensive. Such ruble comparisons, therefore, exaggerate the level of U.S. activities, which are capital-intensive relative to the Soviet level -- the reverse of the distortion that occurs in dollar comparisons.

Whether measured in dollars or in rubles, Soviet defense costs exceeded U.S. spending by a considerable margin in the late 1970s. Total Soviet defense costs in rubles were 30% greater than those of the United States in 1979. Soviet defense costs measured in dollars were 50% greater. That is, the Soviet-to-U.S. ratio of total defense costs measured in dollars was 1.15 times the ratio measured in rubles.

Modeling Problems

Aside from the index-number problem, a number of other criticisms have been made of the direct cost modeling methodology.† What particularly brought these issues to light was the fact that in 1976 the intelligence community doubled its ruble estimate of Soviet defense expenditures from 6 to 12% of the Soviet gross national product.††

The principal criticisms deal with:

- Making an accurate and complete estimate of Soviet military production.

- Accounting for qualitative improvements that are made on existing Soviet weapon systems.

- Accounting for the more complex and higher-quality U.S. technology when using U.S. methods in estimating costs of Soviet weapons.

- Properly estimating the technical qualifications of Soviet military personnel when valuing Soviet manpower at high U.S. pay rates.

III. CONCLUDING REMARKS

The use of models in national policy analysis has grown substantially since the 1960s. Arms control and the estimation of Soviet defense expenditures have been two of many areas of application. Models should also become useful tools in a number of other policy areas in the future.

---

†See testimony before the Subcommittee on Oversight of the Permanent Select Committee on Intelligence, House of Representatives, September 3, 1980. Also, Steven Rosefielde, False Science: Understanding the Soviet Arms Buildup. An Appraisal of the CIA's Direct Costing Effort, New Brunswick, 1982.

††The dollar estimate was essentially unchanged. The ruble change was made on the basis of covert intelligence that was obtained. The primary reason for the discrepancy between the estimates made before 1976 and the revised estimates was that the intelligence community did not realize the extent to which military production was subsidized in the Soviet Union until 1967. In 1967, the Soviet Union undertook a major price reform. One of the things which was done as part of this reform was to remove substantial subsidies on military procurement.

Four key points which apply to all modeling are especially noteworthy in policy analysis. These points were well-summarized in a report† to Congress by the Comptroller General:

- Quantitative decision making is beneficial only when it embodies, rather than replaces, expert judgment and "objective" fact.

- Analyses may give the appearance of scientific work but may not have been subjected to the normal evaluative standards of science.

- The theory and supporting data may not equal the quality of the analytic tool.

- The assumptions and limitations of the analysis must be made a part of the study report.

REFERENCES

1. *Bruce W. Bennett, Assessing the Capabilities of Strategic Nuclear Forces: The Limits of Current Method, The Rand Corporation, June 1980.*

2. *Gilbert C. Binninger, James T. Powers, and John A. Shannon, Examination of Soviet Perceptions of Nuclear Weapons Effects (U), Science Applications, Inc., July 1981. SECRET*

3. *James L. Foster and Bruce W. Bennett, Assessing the Strategic Competition: Methods, Measures, and Matchups (U), The Rand Corporation, December 1980. SECRET*

4. *Thomas L. Saaty, Mathematical Models of Arms Control and Disarmament, Wiley, New York, 1968.*

5. *Models, Data, and War: A Critique of the Foundation for Defense Analyses, GAO, March 12, 1980.*

---

†*Models, Data, and War: A Critique of the Foundation for Defense Analyses, 1980.*

ALFRED LIEBERMAN

Alfred Lieberman is the Chief of the Operations Analysis Division of the U.S. Arms Control and Disarmament Agency. In this position, he is the Senior Advisor to the Director and other officials of the Agency on issues involving quantitative analyses and computer support. He has been with ACDA since 1963, and for 10 years he headed the Office of Operations Analysis. His principal areas of work at ACDA have been in the quantitative analyses of strategic arms limitations (SALT, START, and INF); nuclear test limitations (CTB and TTBT); and verification. Previous positions include Senior Analyst at the Institute for Defense Analyses (IDA) from 1958 to 1961, and Mathematical Statistician at the Bureau of Ships, U.S. Navy, from 1956 to 1958. Mr. Lieberman has served on the MORS Executive Council and Board of Directors.

ABOUT CHAPTER 11

To our knowledge, this chapter by Marion Bryson is the first attempt to describe in any detail the characteristics of field exercises and experiments as models. The chapter goes on to compare the advantages and disadvantages of field experiments with other kinds of simulation, including a carefully considered assessment of some of their less obvious limitations.

Because of Bryson's long experience in constructing and evaluating ground combat field tests, this chapter has a distinct Army and Marine Corps flavor, a taste of the Air Force, and little on Navy fleet exercises and experiments. However, we believe liberal extrapolations from his general conclusions are entirely appropriate for all kinds of field tests and experiments.

Experiments
by
Marion R. Bryson

The Continuum of Models

Analysts have simulated military actions along a continuum from purely symbolic representation to, as nearly as possible, a physical portrayal of the action. At one end of this scale of simulation are the purely symbolic and rigorously mathematical representations expressed in the form of interlocking algorithms. At the other end are empirical representations in the form of experiments. Like symbolic models, experiments are designed for the purpose of deriving relationships between influencing factors and response variables. Unlike symbolic models, experiments portray the influencing variables and players physically.

Field or *fleet experiments* can be defined as combat simulations with the following characteristics:

a. Actual physical representation of opposing forces with conflicting combat objectives.

b. Setting in the actual or analogous environment.

c. Inclusion of trained military personnel.

d. Use of actual or surrogate equipment.

e. Involvement of a control mechanism (umpires) to enforce the rules, and a data collection mechanism (instrumentation). A computer may be an integral part of this control mechanism.

The experimenter processes and analyzes the collected data, just as the symbolic modeler does, to derive relationships between "causes" and "effects." Relationships in both cases are expressed in the form of equations, formulae, graphs, distributions, limits, or other statements.

A *field test* may be thought of as a special case of an experiment in which the purpose is to compare empirical results to a specified threshold value. The threshold for comparison may be a prestated standard or it may be an observed result of the baseline portion of the test. In either case, a field test has a "pass or fail" implication, which makes it a subcategory of the general class of experimentation, since an experiment may also be for the purpose of discovering relationships without the involvement of thresholds.

Those *field* or *fleet exercises* which have as their purpose to generate analytic data may be considered a special case of experimentation. (Exercises may be for non-data purposes as well, such as training.) When the intent of the exercise is to produce data, the results are usually measured in terms of force movement; command, control, and communications; efficiency of intelligence acquisition and use; deployment strategies; maneuver or task force movement; logistics support; and state of training. Since results of exercises are usually measured in terms of development of overall situations, the results are often largely qualitative. Military exercises can be thought of as war games with real players, equipment, and environment.

Unless it is a test or exercise, an experiment typically consists of smaller-size forces operating in a bounded area, with training level established before the start of data collection. Theater-scale experiments rarely occur and never comply with all of the attributes of an experiment hereafter described. The most rigorously controlled field or fleet experiments yield largely quantitative results in terms of target detection, recognition, and identification; firing results; and casualty assessment. A field experiment may be thought of as a small-scale, high-resolution stochastic model with real personnel, equipment, and environment.

The remainder of this section will deal with the field experiment as it has been defined above.

Uses of the Field Experiment

The versatility of the field experiment as a tool for the analyst is great. To get an idea of its possible uses, one need only to think for a moment about answers to a question first posed by Mr. David Hardison, Director, US Army Concepts Analysis Agency: "What is it that I would like to know, that I don't know, that I can find out from a field experiment?"

One use of the experiment is to provide empirical data input to symbolic models. Since an experiment is a simulation in which human and equipment behavior are observed and measured rather than modeled, distributions of those measures are necessary for generating the appropriate stochastic processes present in the symbolic model. For example, in modeling the detection process one needs to know the dependence of the detection time on such things as range, terrain, target size, target movement, and the like. In the experiment the detection process is natural, so the measures taken in the experiment are used to create empirically generated inputs to a computer model of the process.

A second use of the experiment is to validate symbolic models by assisting in rejecting, modifying, or corroborating other models of the combat process. The overall results of several replications of a field experiment and the outcome of several runs of a symbolic model on a computer should be similar. Comparisons may be made between kill rates, movement rates, battle times, intensity of battle, and detection rates. Large differences between the outcome of the computer model and the experiment, such as when the two distributions are non-overlapping, may be investigated in detail to determine the cause of the divergence. These investigations will often reveal improper procedures in one or both methods of simulation.

The above descriptions of two uses of experiments address their role as a combat model in support of other less empirical models. In addition to its support role, the experiment has many uses in a "stand alone" mode. The experiment is useful to the combat developer for designing tactics and techniques of employment of new equipment (or new uses for old equipment). It is useful to the trainer in developing new and better training methods and measuring the effectiveness of each of these methods. It can be used to measure and compare the relative effectiveness of two or more ways of organizing a combat unit.

And it can be used to measure the effectiveness of conceptual or prototype weapon systems. An experiment should be conducted early in the conceptual phase of the materiel acquisition cycle to assist in defining the desired operational capability of a new piece of equipment.

Uses of the Computer in Field Experiments

The function of the human umpire in an experiment is to assure that the simulation is run properly and that no more artificialities than necessary creep into the trial. The trial should be conducted under conditions as close to actual combat as possible. The instrumentation installed on the players should be "invisible"; that is, it should not affect the way a player performs his role in the trial. He should not be part of the instrumentation system in that he should not be asked to punch a button or turn on a camera or report an event on the radio or perform any other duty that he would not normally perform in combat. The umpire does all of those things. The umpire must also be alert to assure that the player does not use the instrumentation to his advantage artificially; for example, the player may disengage the fire cue sensor, thereby making himself immune to enemy fire.

The master umpire in an experiment on an instrumented range is the computer. It monitors the battle continuously, keeping track of all of the players and their activities. It simulates capabilities in the players which they should have but may not possess. But most importantly, as an umpire the computer measures the appropriate circumstances surrounding an engagement, determines the probability of damage or death to the target, and by a random process assesses whether or not the target is a casualty. If the target is a casualty, the computer notifies the target of this fact and prevents that player from firing for the rest of the trial by disabling his weapons. An umpire detonates a smoke cue to inform other players of the "kill" and assures that the dead target takes no further action.

A second major function of the computer is to gather, sort, and store the data generated in the trial. The data gathered and stored should be of sufficient quantity and quality that the simulated battle, as it actually occurred, can be reconstructed after the trial. The computer is then used off-line to analyze the data.

A third important use of the computer is to generate a visual display of the battle on a cathode ray tube or screen in real time. Stored in the computer is a digitized representation of the terrain over which the trial is run. This representation, in the form of contour lines, is shown on the visual display. The location and actions of the players as determined in real time are then superimposed on the terrain display. This display is used by live umpires to detect improper or ineffective tactics. These displays may be used without the terrain representation and may be in three dimensions. One can then view the battle from any angle, vertical or horizontal. The data which drive the visual display are stored in the memory of the computer so that the battle can be replayed at a subsequent time. In either real-time or post-trial play the operator has his choice of scale and in post-trial also a choice of speed of the action, say five times real time. It would be nice to have this last capability in real time, but alas our machines are not yet precognitive.

Corroboration of Results of Field Experiments

One of the uses of experiments, described above, is the corroboration of symbolic models. The experiment, being closer to combat, is more likely to be a faithful reproduction of what would happen in combat under similar conditions. Nonetheless, the experimenter is not relieved of the task of corroborating his experimental results.

We will discuss six ways of corroborating a field experiment:

a. Comparison with combat history. Like the computer modeler who appeals to the more realistic field experiment for corroboration, the experimenter may appeal to the history of like combat actions for corroboration. The senior military officer has been in combat; he has seen the results of military actions. With a given set of circumstances, he can postulate a spectrum of possible outcomes with which to compare the outcome of the experiment.

b. Military judgment. If the outcome of a field experiment does not agree with the intuitive reasoning of the experienced military leader, the validity of the experiment may be open to question. Under these circumstances, the experimenter must be able to demonstrate to the officer why the result was counterintuitive.

c. Internal consistency. In comparing an experiment with itself or (stated another way) comparing the results of several replications of the same experiment, one is not so much testing the validity of the experiment as its reliability. If the same outcome occurs again and again, we have great reproducibility but may be reproducing the wrong answer reliably. Nevertheless, reliability is a desirable characteristic of experimental results; otherwise too many replications are required to reduce the variance of the mean outcome to an acceptable level.

d. Comparison with other experiments. Field experiments conducted by US forces as well as by allied military establishments are often used for comparison. Both the British and the West Germans have conducted experiments which are similar to those done by the US. If two experiments designed and conducted by two different nations but with similar objectives have similar outcomes, that is reasonable corroboration.

e. Comparison with other models. As alluded to earlier, comparison of field experimental results with the results of war games and symbolic models not only helps corroborate the war games and models but helps corroborate the experiment as well.

f. Realistic representation of independent variables. We all operate on the assumption that if we faithfully construct a realistic portrayal of forces and environment, then we can believe that the results portray a valid set of responses. In fact, if we have sufficient confidence in our structuring of the influencing factors, we might accept the results over conflicting evidence from the other five corroborative processes.

Realism of Field Experiments

As one goes from the most abstract combat simulation, the analytic model, to the most realistic, the experiment, the understanding of simulation increases. In the experiment, the most observable simulation, by occupying a well-chosen vantage point one can essentially observe the entire battle. He can see the movement of the forces. He can observe the opportunities to engage and the number of these opportunities taken. He can see elements breaking tactical formation or getting lost. In short, he can see why things are happening as they are. This gives the military leader an opportunity to determine the degree of realism in representing the experimental factors, such as environment and events, and therefore determine his degree of confidence in the validity of the experiment and make modifications for greater realism based on military experience. This is an opportunity generally denied to the leader in computer model simulations.

Even with experiments there are simulated procedures which cannot be physically observed. All of the weapon effects are simulated in the computer much the same as they are in other models. For example, the determination of the probability of kill for an engagement is an abstraction based upon such things as salvo equations. This process is not generally fully understood by the non-analyst.

Problems and Limitations of Field Experiments

Up to this point, the reader may have concluded that experiments are a panacea -- that they do everything better than other methods of studying the combat process. Unfortunately there are some aspects of experimentation which reduce its usefulness. The most obvious disadvantage of the experiment is the cost per unit of information. Determining the cost of an experiment is an imprecise operation at best, but the cost of one replication of an experiment can be orders of magnitude greater than one replication of a model. Because of the high cost in money, time, and equipment, only a very few points in a sample space can be investigated by experiments. From those investigated points, some extrapolation and interpolation are possible. For example, we routinely extrapolate in time and space from the results of an experiment. Unless we believe that the results of a field experiment have validity as a predictor of what will happen if the combat takes place at a later time at another place, the value of the experiment is essentially zero. It follows that another limitation or pitfall of the experiment is the extrapolation of the results to situations in which they do not apply. The effectiveness of a weapons system as measured in the mountains of California should not be interpreted to mean that the system would be equally effective in the plains of West Texas. The effectiveness of an aircraft system employed against a force using a 1975 vintage enemy air defense system cannot be extrapolated to a similar effectiveness against a 1990 air defense system.

Another limitation of field experiments -- indeed a limitation to some extent of all combat simulations -- is the play of the opposing force tactics and equipment. Because we do not have better information, we often play our conceptual or prototype future equipment against the outmoded tactics and old equipment of the opposing force. Are we not analyzing the two opposing forces with a complete disconnect in the time frame?

The artificialities present in a field experiment undoubtedly have an effect upon the behavior of the players. Specifically, and most importantly, the absence of danger of physical injury or death as a result of enemy actions must have a large but unknown effect. Such well-known combat phenomena as fear, panic, heroism, stress, and battle fatigue are absent and can be simulated only with great difficulty if at all. It is quite likely that in some cases player behavior in combat can be better simulated by a symbolic model than by players in an experiment.

Solutions? Cost we must live with. Judicious choice of points in the sample space can gain the greatest information per dollar. Instead of extrapolating analytically from the points investigated experimentally, the stochastic model, corroborated against the points in the experiment, is a far better way to investigate the response surface.

For example, if one is investigating the effectiveness of equal cost forces with different weapon mixes, a field experiment may be used to measure the effectiveness of, say, three candidate mixes. A stochastic model can be designed to be a computer representation of the field experiment. Using the model, hundreds of trials of the field experiment can be simulated to investigate many other mixes. The results of the actual experiment are used to "anchor" the model at those three points. The problem of playing the wrong enemy force is much more difficult. One way to gain some insight into the effectiveness of our future weapons is to play our future offensive weapons against our own future defensive weapons. It is highly likely that the 1990 enemy air defense weapons will have a capability more like *our* 1990 air defense weapons than like *their* 1975 air defense weapons. Another way to design the opposing force is to appoint a red team who will determine the weapon capabilities and tactics to be played in the experiment. In the experiment, one can allow the red force to learn and adjust just as the blue force does. Models don't generally allow for learning to carry over from one battle iteration to the next; experiments do. This is a mixed blessing, however, because some of the learning is how to fight *that* battle rather than how to fight *a* battle. An example of how learning can incorrectly affect an experiment is a practice called "pinballing." If early in an experimental trial one side does especially well, that side may disengage or play very conservatively to avoid making a tactical error which would invalidate that trial.

The solution to the problem of lack of combat stress is the most difficult of all. Attempts are made to substitute the notions of competition and peer approval for the stress of actual combat. The degree to which these attempts are successful is open to question. This is a deficiency with which we live.

Another problem peculiar to field experiments is caused by their transparency and their high trial-to-trial variability. A high-ranking individual, observing but a single replication, may draw unwarranted conclusions based solely on what he saw. What he saw as a defect or as a strong point of some military system may have occurred only in that trial. In the analysis of all the results, such an isolated occurrence would have little effect; to the high-ranking officer, it is much more important.

Believers and Non-Believers

As with models in general, whether one is a believer or a non-believer in the results of an experiment is highly correlated with whether or not the results are supportive of that person's point of view. It is a simple fact of human nature that one is more likely to believe the contest is fair if he wins than if he loses. So we should borrow from the sports world, assume that there are advocates and adversaries for every investigative outcome, and plan accordingly. There are three actions necessary: design fair rules and assure that all understand and accept the rules. Because of their realism, experiments have an advantage relative to understanding the rules. More of the combat processes are played naturally; hence more are understandable. Because of this understanding, acceptance of the rules can be more difficult than it is for models. The adversary, as well as the advocate, can more easily find ways in which the rules are detrimental to his point of view. But more often than not, this discovery of "bad rules" is a result of Monday morning quarterbacking rather than a result of pretest study of the rules.

Experiments share this aspect of human nature with all other types of models. However, there appears to be a difference as celebrated in the old saw: "No one believes the results of a model except the modeler; everyone believes the results of a test except the tester." This seems to be a graphic overstatement of both the greatest strength and greatest weakness of the experiment's place in the continuum of military operations research models. It recognizes that the experiment yields a much more empirical result from a much more transparent representation than the symbolic model. It also recognizes that the experimenter is always painfully aware that field data are much cruder and much less reproducible than symbolic models. The correct position for the military analyst is probably to keep the experiment in the proper perspective among the family of modeling techniques, capitalizing on the strengths of all types of models.

DR. MARION R. BRYSON

Marion R. Bryson received a Ph.D. in Statistics from Iowa State University in 1958. He divided his time during the next 10 years between teaching in the Mathematics Department at Duke University and working with the Army Research Office in operations research. For four years he was the Technical Director of the CDC Systems Analysis Group. From 1972 to 1983 he was the Scientific Advisor, and in March 1983 became the first civilian Director of the Army Combat Developments Experimentation Command, Ft. Ord, California. He has published numerous papers on the development of field test methodology and the analysis of test data. He holds membership in several professional organizations and is a past President of MORS.

ABOUT CHAPTER 12

As the Overview says, an insideous mistake made even by experienced policy makers is to adapt models and their results for purposes never intended by the analyst. The problem goes far deeper than a simple misuse of data or scenario, which is obvious and quickly caught, but gets into the model structure itself, which is subtle and often escapes attention. How and why model design affects model results is, of course, a central theme of this book. Nevertheless, the subject is important enough that we have devoted this short chapter specifically to hazards of model misappropriation and ways to avoid them.

John Battilega and Judith Grange have drawn from their four-volume research report, *The Military Applications of Modeling* (1978). This chapter is an adaptation from one section of it.

The Misappropriation of Models
by
John A. Battilega and Judith K. Grange

INTRODUCTION

During the past twenty years, many computer models have been designed, developed, and delivered for use by the Government and industry to support military planning. Some of these models look duplicative because they focus on similar forms of military operations. For example, the annual *Catalog of War Gaming and Military Simulation Models* published by the JCS Studies Analysis and Gaming Agency routinely lists many models focused on land warfare, air warfare, naval warfare, strategic nuclear forces, and so forth.

Because of the number of existing models which treat similar areas of military operations, a decision must be made regarding which model or models should be used to support a particular study. This decision is normally based on such factors as model access, analyst experience in the preparation of input data, the prior familiarity of the analysts or study directors with a model by virtue of previous studies, and bureaucratic negotiations among study participants.

Unfortunately, model selection decisions sometimes have unintended (and even unsuspected) consequences. Conclusions drawn from the analysis of model results may be inappropriate if the study team does not have rather complete knowledge about the process which led to the design and development of a particular model in the first place. Design teams and modelers, confronted with the requirement to develop a new model, must make several technical choices. For example, they must precisely select those aspects of the given warfare area they are going to model in detail and those aspects which they are going to approximate. They must decide what internal decision logic is going to be explicitly incorporated into the model, including the specific quantitative criteria which are going to be used in the model as the basis for decision. They must decide what class of decisions the model is intended to support. They must be sure that the critical uncertainties unique to that class of decisions are properly reflected in the model. And finally, they must decide the form and content of model output which is consistent with the overall model design objectives.

The results of such technical decisions produce a model which is internally consistent with respect to specific categories of analysis, and which has incorporated proper trade-offs with respect to the breadth of issues and the level of technical detail associated with that analysis. Collectively, the relationships between the kinds of technical decisions made by modelers to produce this consistency, the warfare area being modeled, and the classes of studies for which the model was originally designed comprise the *structural assumptions* of a given model.

Sometimes models are adopted for purposes which violate those structural assumptions. In this sense, the model is misappropriated for uses other than that for which it was designed. Sometimes such misappropriations lead to contradictory study results; at other times, they provide results which are technically invalid. At best, such misapplications of models provide eventual grief for the study teams which use them. At worst, they lead to poor decisions. This chapter briefly discusses some of the major model design principles which typically are violated when models are misappropriated.

SOME MODEL DESIGN PRINCIPLES AND THEIR IMPLICATIONS

1. Modeling the Analysis Criterion.

In some cases, the primary decision criterion incorporated into a model is not a "victory" criterion appropriate for a combat situation, but rather a measure of effectiveness (MOE) focused on the decisions the model is designed to support. The use of the same model for studies based on other analysis criteria (i.e., different MOE) may produce invalid results. For example, studies used to assess alternative nuclear weapon systems which can inflict damage on a prioritized set of targets have led to the design of models in which the criterion used to decide how to assign weapons to targets is the mathematical expectation of the damage which would result to that set of targets, weighted according to the "value" attributed to the targets. To use that same model when the analysis criterion is based on the number of targets attacked, or the number of targets of a given kind (e.g., missile launchers) which will be destroyed by the attack, is a misappropriation of the model. The mathematical equations involved in the design of a targeting plan which optimizes results according to either of these latter criteria have a different form than that for the former. Even though these latter kinds of statistics are normally available as model outputs, the optimization which generated the allocation was based on a different criterion.

2. Use of a Single Decision Criterion.

A single model is usually developed around only one internal primary decision criterion. This inflexibility is normally dictated by the difficulty of orienting internal decision logic to the simultaneous treatment of multiple measures. A misappropriation may occur when studies are based on secondary decision criteria. Recently, the use of multiple measures of effectiveness as the basis for decision has been an area of research, and some of the newer models attempt to incorporate such considerations. Even these new developments, however, typically require the analyst to specify relative priorities or weighting among criteria. The internal logic of the model then mathematically combines these priorities into an artificial decision criterion which is single-valued and oriented toward the model builder's decisions of interest.

3. Spectrum of Methods for Modeling Uncertainties.

Uncertainties which comprise key elements of analysis are dealt with in model design by several different methods. The methods include explicit modeling, implicit modeling, parametric modeling, and hedging. Both the intended use of the analysis and the availability of data strongly affect how a modeler will choose between these quite different methods of modeling uncertainty. In some cases, the modeler deliberately assumes that the uncertainties will be dealt with parametrically by input variations; in other cases, the uncertainty forms part of the internal model logic. In still other cases, the uncertainty is modeled explicitly as a random variable. For example, the uncertainty in weapon impact point caused by guidance errors is normally dealt with explicitly via the specification and integration of probability distributions. The uncertainty caused by lack of precise knowledge about the actual scenario for nuclear war is dealt with implicitly by employing canonical bounding scenarios such as the assured destruction scenario. The uncertainty caused by lack of precise knowledge about strategic bomber penetration probabilities is frequently dealt with during the analysis by varying the penetration fraction parametrically in a model which has been designed to accept a single numerical input value for that fraction. The uncertainty caused by the possibility of unanticipated failure of one or more legs of the strategic TRIAD is treated by hedging the targeting plan to require cross-targeting or by specifying separate and redundant targeting goals for each leg of the TRIAD separately. Because of these vastly different ways to handle uncertainties in the design of a model, the treatment of key uncertainties in any study must find precise reflection in the structural assumptions of the model used to support the study. Any mismatch between the model and the treatment of uncertainties would be a misappropriation.

4. Design for Specific Breadth Versus Level of Detail.

Modeling can be in great breadth (e.g., massive theater-force campaigns) or great detail (e.g., blast effect on bridge trusses), but not at the same time. Great breadth and detail simultaneously not only will normally violate the cost, time, and computer resources of a given study, but generally will also exceed the availability of data and the ability of the analyst to interpret the results. A model designed to treat broad issues (e.g., the appropriate mix of armor, infantry, and artillery in a land campaign) is generally misappropriated if used to study detailed interactions in that same warfare area (e.g., alternative single-tank tactics). Conversely, the use of a very detailed model to study a very broad set of issues is unworkable in any study of realistic scope. Moreover, the internal logic of a detailed model may preclude the study of many issues which may be important in studies of great breadth. This occurs when, for example, the detailed model is based on an assumed specific resolution of the broad issue. (For example, a detailed model of mobilization demands that large issues related to the course of a campaign as a whole be fixed in order to reduce the problem to manageable size for the purpose of modeling.) It would be a misappropriation to use a model in a study which exceeds either the breadth of analysis or the level of detail for which the model was designed.

5. Model Design Based on Analytic Scenarios.

Analytic (rather than "real" or "most likely") scenarios are frequently invoked as the basis for the design of models in order to allow the study of difficult problems at the boundary of major uncertainties. For example, a massive counterforce attack is frequently built into models designed to study the capability of the land-based ICBM force to fulfill an appropriate retaliatory mission. It is a misappropriation to use such models to study the probable evolution of general war campaigns involving the employment of intercontinental nuclear forces.

6. Deliberate Offensive or Defensive Biases.

Modeling assumptions are sometimes deliberately biased pessimistically with respect to either the offense or the defense. For example, a model of aircraft penetration, used to size bomber forces for strategic missions which require a high confidence of execution, may deliberately attribute knowledge to the defense which results in pessimistic (lower than probable) penetration estimates. The use of that same code for analysis of the defense options of the opponent against a constant attack force would overestimate the effectiveness of the defense. Furthermore, depending on the specific model, such a bias may not be the same for all candidate defenses. Hence the offense-designed model would be misappropriated when applied to defense-oriented studies.

7. Implicit Biases in Favor of Particular Weapon Systems.

Models occasionally are constructed which deliberately focus on one or two technical details or aspects of combat operations. Implicitly they will emphasize the favorable characteristics of specific combat systems to the detriment of others. (Sometimes this emphasis is unintentional and results from the representation of different elements of the situation in different degrees of detail.)

Modeling aggregation then can cause a bias toward certain sets of weapon systems. For example, a multipurpose warship will compare unfavorably with a single-purpose ASW ship in a model in which the technical aspects of ASW performance are given special emphasis.

8. Structural Linkages Between the Model and the Data.

The "model" and the "data" that support the model are almost always structurally related. For example, models of current weapon systems frequently may contain modeling assumptions based on current technical parameter ranges. Using the model to study qualitatively different systems is a misappropriation which may produce unintended and invalid results. This can easily occur when models based on the characteristics and doctrine for current forces are used to study new weapon systems, force postures, or operating concepts for combat operations set many years in the future. Another related difficulty can arise because data bases typically are maintained for specific models and sometimes are partly hard-wired into the software program. These data bases contain numerical values (e.g., target acquisition probabilities) which are typically based on certain assumptions about a variety of combat parameters. For a particular study, these "standard" data bases must be carefully reviewed to insure that the campaign assumptions of the study match those under which the data was generated. It typically is not sufficient to review the table of inputs; input values hidden inside the program must also be sought out for consistency.

9. Modeling Specific Planning Situations.

Models are normally developed to give quantitative visibility to a specific set of military alternatives in a reasonably well-defined military planning situation. Problems inevitably occur when such a model is used to support studies which do not focus on that well-defined situation. Most notably, the use of a force planning model for operational planning will generally produce unsatisfactory results. The internal assumptions of a force planning model are consistent with the many uncertainties associated with future conflicts and are tailored to support broad decisions on the allocation of resources for force improvement. As a result, force planning models do not give sufficient attention to major details which are critical for development of operational plans for current forces. The use of a force planning model to study the employment of current forces generally is a misappropriation.

10. Calculating Criteria Versus Simulating Combat.

Many times models are developed to calculate a specific measure of effectiveness (MOE) under well-defined assumptions, rather than to "simulate" a combat situation. This usage most typically occurs when neither historical information nor direct experience about campaign dynamics exist in sufficient detail to provide the basis for a simulation. Alternatively, the model may be used for a purpose only indirectly related to combat outcomes. The first situation is typified by models that focus on intercontinental nuclear warfare; an example of the second is the calculation of the minimum total yield of a land-based missile force which could be expected to survive a preemptive nuclear attack. (In the latter case, the yield or its equivalent is a force-level aggregate typically used to support nuclear policy debates.) Models designed to calculate an analytic measure of effectiveness have structural assumptions consistent with that purpose (as opposed to, for example, giving visibility to the most important parameters of a combat situation). Use of a model of this type for purposes not directly related to the measure of effectiveness will in all likelihood give misleading results.

11. Relative Versus Absolute Values.

Some models are intended to estimate the absolute value of combat parameters; other models are intended only to study the relative values of parameters. For example, models exist to provide the best estimate of the absolute number of men and pieces of equipment in an infantry battalion which will become unable to function in combat as a result of a nuclear attack. On the other hand, models also exist which are designed solely to provide particular quantitative insights into the *relative* value of fixed versus mobile ICBM basing for strategic nuclear warfare. It would be a misappropriation to use models in the latter category for the purpose of developing a realistic estimate of the actual number of ICBMs (in either basing mode) which would survive a realistic nuclear attack as part of a general war.

12. Techniques Providing Invisibility to Certain Issues.

Some very useful analytic or modeling techniques, once embodied in a model, will preclude the analysis of certain kinds of issues. For example, certain theater ground combat models that relate "expected" casualty rates, movement rates, and force ratios cannot show the effects of breakthroughs in specific sectors of the battlefront. (For forces of comparable size, breakthroughs occur because of the rapid development and exploitation of other-than-expected situations.) It would be a misappropriation to use such models to study the relationship between one's force mix in a theater and the likelihood that the force mix would be sufficient to prevent a breakthrough.

13. Deterministic Versus Expected Value Modeling.

"Expected value" models may be neither "most likely" nor "typical." Depending on the nature of the input data, they may also not really be "expected value." In fact, "expected value" modeling is more properly called deterministic modeling, even though generally some of the processes (and some of the parameters) are indeed the mathematical expectations of the random variables involved. Few, if any, models, however, actually calculate the mathematical expectation of the total set of random variables implied by all of the combat processes being modeled. Even parameters originally modeled correctly as expected values sometimes become simply model input controls used by analysts; and in this case, their interpretation as expected values is questionable. In general, it is wrong to assume in a study that an "expected value" model will give "expected" outcomes (in a mathematical sense) of the combat situations being modeled.

IMPLICATIONS

The wealth of variations in structural assumptions underlying models create situations where many different models exist to study the same phenomena (such as land combat). The differences derive from the fact that many different types of *decisions* must be made in order to design, develop, test, assess, and operate a military force. As a result, there is a requirement for families of models, each designed to support specific types of studies. In each category of combat, the input and output variables for several of the models may have strong similarities. The algorithms that are used to process the inputs and produce the outputs, however, may be quite different. Each may be valid for the applications for which they were designed -- yet paradoxically they will frequently not produce the same results if run with the same inputs. Bemused clients have every right to demand an explanation when different model results both purport to apply to the same *decision*. Frequently the explanation will be that a model was misappropriated.

No standard way exists to characterize the total set of structural assumptions which are associated with any given model, nor are there standard methods for assessing the significance of deviations from the requirements of a specific study. Further, once the members of an analysis staff become familiar with the inputs and outputs of a computer program, the structural assumptions often tend to be taken for granted and cease to be scrutinized for specific applications. In fact, after a couple of generations of analysts have passed, the institutional memories of the organizations are often inadequate to support such scrutiny.

Ultimately, the responsibility for avoiding misapplication must rest with the study team. The most critical step in the study must be the development of the list of internal assumptions *of the study*. That is, the analysis criterion should be stated explicitly, appropriate decision criteria should be identified, the nature of the uncertainties should be characterized, the appropriate levels of breadth and detail should be stated, etc. This process, when executed rigorously, will provide a very detailed picture of the calculations that need to be performed in order to do the analysis. The applicability of available computer programs can then be judged. (This process of designing the study and then performing selected calculations contrasts with a not unusual process of performing calculations first and then examining the results for relevance to the objective of the study.) In some cases, standards for documentation of the computer programs, which tend to describe the algorithms and their implementation in computer languages but ignore the basis for the selection of the algorithms, may need to be altered so the structural assumptions for the models are more clear.

DR. JOHN A. BATTILEGA

Dr. John A. Battilega holds a Ph.D. in Applied Mathematics from Oregon State University. As an Army officer he served with the 101st Airborne Division in Vietnam, the 6th Armored Cavalry Regiment, and the Strategy and Tactics Analysis Group. He has spent the last twenty years in military operations research, concentrating on nuclear force planning, arms control, and intelligence. He is also a specialist in Soviet military operations research. He has been on the technical staff of the Martin-Marietta Corporation, and for the last ten years has been with Science Applications, Inc. (SAI), directing the development of many computer models in widespread use throughout the Government. For the last five years he has served as Vice-President and Director of SAI's Foreign Systems Research Center. He is on the Board of Directors of MORS, and has coedited a textbook on <u>The Military Applications of Modeling</u>.

JUDITH K. GRANGE

Mrs. Judith K. Grange holds a Master's Degree in Mineral Economics (Operations Research) and a Bachelor's Degree in Chemical and Petroleum Engineering from the Colorado School of Mines. She has worked for Standard Oil, Amoco, and Stearns Rogers as a design engineer. For the last seven years she has been employed by Science Applications, Inc. (SAI) as a military operations research analyst, concentrating on nuclear force planning, arms control, and intelligence. She is a specialist in Soviet military operations research, and is currently Assistant Vice-President and Deputy Director of SAI's Foreign Systems Research Center. She is a consultant on Soviet military planning to many Government agencies and is a lecturer at several military schools. She has coedited a textbook on <u>The Military Applications of Modeling</u>.

ABOUT CHAPTER 13

The next two chapters discuss model verification and validation -- two terms that have been used so interchangeably that the collaborators in this monograph have not undertaken to disentangle them.

In this chapter, Clayton Thomas will define validation as the process of determining the correctness of a model's internal logic. He shows that even this relatively mechanical validation process can be lengthy and expensive, and not without room for residual error.

Thomas then turns to the problem of verification: the question of determining how adequately a model approximates reality. Lucidly -- yet with considerable thoroughness -- he discusses four problems that bedevil not only operations analysts but scientists in general. The first is to establish the basis of "reality" itself. The next three he identifies memorably as the "teleological problem," the "epistemological problem," and the "uncertainty principle" of modeling. We will leave an understanding of the nature of these difficulties to the reading of Chapter 13. Suffice to say, the reader will come away with a deeper appreciation of the provisionality of all model verification.

Still, Thomas's central conclusion is that problems of validation and verification simply imply a greater need to struggle for excellence. It is a conclusion we all would echo.

Verification Revisited -- 1983

by

Clayton J. Thomas

Introduction

In 1969, the Center for Naval Analyses (CNA) published a transcript of a Warfare Model Verification Conference held at CNA on October 1, 1968. As one purpose, CNA hoped to solicit from the conferees thoughts on model verification†, especially the use of combat data. CNA was considering a possible study of the interaction between air forces and ground combat forces that would depend on the use of a particular model, and hoped to test that model against some Korean War data. It was thought that the experiences of the conferees might suggest useful procedures. In a broader sense, however, the transcript indicates that the conference was undertaken because "the subject was ripe for general discussion."

One basis for this judgment was the ongoing Vietnam War. On the one hand, the war provided combat data that might be adapted to yield input values for computer simulation models. On the other hand, however, and more relevant to the verification of those models, the war brought fresh reminders that a new war may differ significantly

---

†This chapter uses "model verification" to refer to an investigation of the agreement of the model with reality, and "model validation" to refer to an investigation of the internal consistency of the model -- the correctness of its logic and mathematics. There is no accepted standard usage in the community, but the definitions of this essay are consistent with much philosophical writing and with the intent implied by the title and purpose of the CNA Conference.

both from previous wars and from our models. These subjects were prominent at the CNA Conference. Over half of the twenty conferees had had experience in Vietnam analysis, and at least a third of the presentations involved the acquisition or use of Vietnam data. Others, similar in spirit, concerned data from the Korean War and World War II.

Very evident in such presentations (and in many other discussions of the same time period) was the theme that the modeling community should take advantage of combat data (and of tests and exercises) to verify models of warfare. Though not as prominent, there were also some important counterthemes. While acknowledging the desirability of verification efforts, it was also essential to note the fundamental limits to verification and therefore to emphasize the value of models even when not fully verified.

The remarks of this conferee were largely in the latter direction, emphasizing general problems of model verification. While those 1968 remarks are not in conflict with my present views, much that has happened in the last fifteen years suggests some shift in emphasis. We see more clearly both partial solutions to old problems and new problems that were only then emerging. It seems appropriate, therefore, to revisit the problem of verification in the light of such developments as larger models, families of models, more institutional models, and trends in the philosophy of science in general.

This discussion begins by noting in the section on "The Enlarged Setting for Model Verification" some of the reasons for a growing concern with verification that emerged in the 1960s and that remains today in the 1980s. The succeeding sections introduce different concepts relating to verification, distinguishing *validation* (as a test of internal consistency) from *verification* (as a test of agreement with something external, through comparison of model results with data, judgment, the results of another model, etc.). Consideration of these concepts then illuminates some of the persistent difficulties that limit the verification of models. The final section concludes, however, with the encouraging observation that our continued concern with model verification -- the verification "process" itself -- brings us reasonably close to goals that would forever elude us if we set our sights too narrowly on individual models or even model hierarchies.

The Enlarged Setting for Model Verification

A reading of the transcript of the CNA Conference shows clearly that the search for verification of models was in large part a quest for credibility. A proximate cause of the conference itself was doubt by CNA management that it should endorse the use of a model in a study unless the model could be shown to produce "reasonable" results. Many decision makers were skeptical of the results of studies that depended on large computer simulations, and rightly so. The analysis community itself expressed similar concerns throughout the 1960s and early 1970s with growing frequency and intensity. One finds examples in the 1967 NATO Conference on Lanchester Theory; the 1968 CNA Conference; addresses of Glenn Kent, Al Flax, and Seth Bonder to 1969 and 1970 Symposia of the Military Operations Research Society; and the 1971 Rand Conference on Models, Games, and Simulations.

The 1960s world of systems analysis and computer simulation models had departed -- with gradually increasing speed -- from the 1940s world of wartime combat operations research. In the setting of World War II, operations research began by bringing the scientific outlook to bear on improving the use of weapons and -- to a lesser extent -- the weapons themselves. There was great interest in getting and using data -- such data as were attainable. This interest is very evident in the sprightly 1943 guidelines of Blackett†, as is also an emphasis on small steps and a distrust of such "a priori" approaches as the use of unverified equations like those introduced by Lanchester. The operations research of the 1940s was embedded in the decision-making process in such a way as to encourage quick results and to underline the provisional nature of any theory.

By the 1960s, however, operations research agencies were addressing the larger problems of systems analysis, and were using new methods such as simulation models implemented on increasingly available electronic computers. Computation in World War II operations research was limited by the resources available. Analysts were often known as "slide rule boys," as slide rules and four-function desk calculators were the common tools. The occasional use of punched-card equipment was notable

---

†*Reprinted in Blackett (1962).*

enough to receive special mention in such accounts of the era as that of Morse and Kimball (1946).

In the early post-war years, computers were still seen as tools to be used in much the same way that slide rules and desk calculators had been used. The changes wrought by increasing speed and size were seen at first as largely changes in degree. They did not at first seem to pose special problems for verification. Even when computers became large enough for programs that combined several sizable subroutines, there was not initially much concern with any new problem of verification. One could argue that such a combination was, in principle, the assembling of separate equations into a system; if each equation is already verified (e.g., on the basis of combat experience), then any difficulty imposed by the combination is largely one of added computational complexity -- a problem for the computer, perhaps, but not a new challenge to verification.

Also, as computer simulation models were introduced, they were at first seen as experimental. They were useful for exploring the potentialities of computers, useful for learning about large-scale systems, and useful in suggesting models. As long as this experimental attitude prevailed -- as long as users saw the true character of a "what if" approach -- questions of verification were less acute and often less relevant than they later became. In many of the early efforts, it was quite clear that an immediate and comprehensive representation of "reality" was not a major goal.

All of these various arguments for not being seriously concerned with verification initially had some merit. As we moved from the 1940s to the 1950s, however, the foundations of the arguments weakened. The combat experience of World War II, partially rediscovered and recreated in Korea, gradually became less and less relevant. As computer programs grew larger and larger, it became more and more difficult to resist the temptation to transfer large models from an experimental role to an institutionalized productive role.

In 1957, for example, what Rand had developed as their Strategic Operations Model was adopted with high hopes by Air Force Headquarters, renamed the "Air Battle Model," and institutionalized in a newly created office that has evolved into what is today the principal agency for studies and analyses of that Headquarters. That transplantation reflected the spirit of the time. All of the Military Services and the large weapons corporations were attracted by the lure of large computerized models. To add to their interest, Secretary McNamara in 1961 created an Office for Systems Analysis that was to require quantitative justification of military programs. To the proponents of such programs there was added incentive to exploit all of the resources of computers in simulation, optimization, and sheer bookkeeping.

Also, however, there was increased incentive to the opponents of such programs, to the proponents of other programs, and to the "just plain skeptical" to question the results of computer programs. Some of the critics were doubtful about the results because they were *not* familiar with computer programs; others were doubtful because they were *all too* familiar with computers. Both sources of doubt led to more concern with methodology and credibility.

In the case of the CNA concern that prompted the 1968 Conference, a single model (TWSP -- Tactical Warfare Simulation Program) was involved, and it was being examined prior to possible use in a particular study. As time went on, however, there was more of a tendency to institutionalize models for repeated application (as in the previously mentioned example of the Air Force and its Air Battle Model). This put even more of a premium on credibility, as such models (termed "standing models" in this monograph) became well-known as representatives of the agencies that maintained them. One increasingly heard questions of the form "Can the ______________ Agency be expected to get reasonable answers, given the kind of models they use?"

Another trend was the introduction of a family of models, sometimes specialized as a "hierarchy," to replace an effort to develop one self-contained and all-purpose model. As an extension of this trend, the development of different members of what is viewed as an integrated family of models may occur in different agencies of one Military Service. Or one may get somewhat the same effect (in an *ad hoc* sense) by borrowing several models for combined use in a given study. In any event, the result of using several models together is to add complexity and difficulty to the already perplexing question of credibility.

The concern with model credibility, which has been with us for practical reasons ever since the use of large models became "big business," has inspired an interest in model verification that goes beyond an idealistic search for truth. Clearly, the argument goes, if a model gets "real" results, it is a more reliable and credible instrument for use in studies and analyses. Thus, verification is seen as one approach to credibility, and this view provides added reason to enlist what is sometimes thought of as the "rigor of the scientific method." Later sections of this essay review some of the difficulties that such a program encounters in even very simple verification efforts. There is special irony in the observation that recent work of such philosophers of science as Thomas Kuhn, Imre Lakatos, and Paul Feyerabend in examining actual scientific *practice* makes it seem very unlikely that one can define or codify scientific *method* short of Feyerabend's "anything goes."

Types of Verification

The above review suggests several principles that one might use in evaluating a model, especially a standing model or family of standing models. Among other qualities, models should be usable and reasonably credible. Each of these qualities has various dimensions, and these dimensions may suggest tests to use in judging models. We need not explore the dimensions of usability (which include run time, preparation time, ease of model maintenance, data base availability, etc.); but credibility, as a basis for concern with validation and verification, deserves more attention here.

If "beauty is in the eye of the beholder," then surely "credibility is in the heart of the believer." Since there are some "true believers" who believe as an act of faith, any complete taxonomy of belief would take us well outside the scope of an inquiry into verification and its problems. Whatever its precise definition, "verification" in normal usage connotes a process aimed at making belief less a matter of faith, not more.

The review of the previous section indicated two types of impediments to credibility that have emerged as principal concerns of model verification. One is the lack of internal consistency in a model, and the other is the lack of "adequate agreement" of a model with "reality."

Removing such impediments clearly seems to be closer in spirit to scientific procedure than to an appeal to faith, however difficult it may be to formulate precise definitions.

It is also difficult to find suitable terminology. For the two notions suggested in the previous paragraph the terms "validation" and "verification" are often used, but not always in the same order. There seems to be no standard usage. In this essay, the term "validation" refers to testing a model for internal consistency, while the term "verification" is reserved for testing the "agreement" of a model with "reality." This usage seems to agree with the notion of "verification" implicit in the title of the CNA Conference.

It may be noted that in a broad approach to credibility, verification (as defined above) would not necessarily imply validation. If, for example, one believed in oracular divination of "reality," then one might not be concerned with validation of internal consistency of methodology. Such a notion, however, seems very remote from what is considered reasonable in the modeling community, and it is difficult to imagine verification efforts that would not involve a concern with validation also. In fact, whatever the practical difficulties of validation, it is likely to pose fewer conceptual difficulties than verification. In that sense, a verification process could very well begin with validation efforts. That is the order followed in this chapter.

Problems of Validation

To validate a model, one must establish with "reasonable" confidence that it is consistent with the principles of logic and mathematics, and that it follows the intent of its outline, flowchart, etc. Testing a model in this sense would not be expected to involve the subtle or profound aspects of logic and mathematics to any great extent. One is more likely to be concerned with correcting such blunders as "probabilities exceeding unity," sums that "don't sum," etc. Validation, moreover, is only the first step in testing the "goodness" of a model, and it remains for verification to examine its agreement with reality.

One might easily be led to think, therefore, that relative to verification, validation is neither very important nor very difficult. After all, validation would seem to rest on a systematic application of generally accepted principles of logic and mathematics, whereas verification poses deep problems for philosophers of science to which there are no agreed answers. One might have expected validation to be rather mechanical and straightforward.

In the face of these observations, it is most striking to note the great magnitude of effort devoted to validation. Measured either by dollars or by rhetoric, there is significant activity -- fully enough to give validation its role as worthy companion to verification. In practice, the validation step seems to be neither trivial nor easy, and it is interesting to note some of the reasons.

First of all, it is important to recognize that a lack of concern with validation would be absolutely unthinkable in a community of modelers. As professionals who seek to invoke the power of the scientific method and the rigor of mathematics, they honor the guidelines and rules of those disciplines. Getting the "right" answer -- important as that is -- must still be done in accordance with methods in the spirit of science, where progress has been closely tied to extending the scope of mathematical models.

Also, as noted, early thoughts on how computers would be used implied a central role for validation. Originally, that is, computers were visualized primarily as tools for the rapid implementation of models already thought to be reliable. Though the ambitions of the modeling community soon went beyond the original modest program, the basic argument for the importance of validation remains: however credible a model may be, a new computer implementation of it requires validation. Moreover, the joining of two or more models, each of which has been separately verified, gives ample opportunity for error and therefore a need for validation (as well as the need for verification of the combination).

Granting the importance of validation, there still remains the question of why it is so difficult. The answer seems to lie in the human tendency to err, especially when a task is long and complicated. Validating a computerized model is much like checking the validity of a long "proof" of a mathematical theorem. Any single step should pose little difficulty. Nevertheless, the lore of mathematics is replete with records of incorrect arguments initially accepted as "proofs." Fermat's "big theorem" provides examples where even very competent mathematicians have gone astray. Such mistakes in mathematical papers are still made on occasion, and despite much care in "refereeing" papers prior to publication, false "proofs" still appear in technical journals.

It is understandable, therefore, that programming errors in computerized models are far more common than one would like, and it is also not surprising that programming experts have given much attention to the detection and avoidance of such errors. During the past decade, they have introduced or standardized a variety of techniques. Some such as "structured programming" and rules for the size of subroutines, etc., pertain to "coding" practices. Some such as certain "team" approaches, "walk-throughs," etc., are devices for "managing" group endeavors effectively. Some such as the use of test cases with known results are efficient diagnostic techniques for the detection of errors.

When one has a family of models, there is an added consideration. Validation then involves not only the testing of each model individually for internal consistency and programming correctness, but also the testing of all relevant interactions of one model with another. If the output of one model is an input to a second model, then the two have to be formally and substantially compatible. This is much more than a question of units (though failure to use common units is a frequent mistake). *Substantial* consistency refers to the treatment of an entity (as, for example, an F-15 airplane) that is represented in several models of the family. If a scheduling model, for example, schedules three F-15 sorties per day, then a maintenance model must involve resources (spare parts, maintenance man-hours, etc.) consistent with that sortie rate. Such a consistency requirement is independent of the verification requirement that a sortie rate produced by a model should agree with a "real" sortie rate.

One sees, therefore, that the validation of a large model, or the validation of a family of models, can introduce a great many problems. The resolution of these problems requires primarily good management, attention to logic, and careful programming. In practice this may involve a great deal of effort, and in the case of a large model (or family of models) there may be no way of ensuring with any feasible amount of effort that validation is complete. Despite these difficulties and frustrations, however, one must be prepared to devote great effort to validation if modeling is to be genuinely scientific in spirit.

Problems of Verification

As defined here, verification involves testing the agreement of a model, or family of models, with "reality." One is required, therefore, to establish what is "reality," to decide on what constitutes "adequate agreement," and to specify what sort of pronouncements are to be made in the respective cases of agreement and non-agreement. Each of these steps, even in the simplest case where one has but a single model to verify, poses fundamental and profound difficulties. This section illustrates a few of these.

To start, what is the standard of "reality?" Is it combat data, as suggested by the statement of purpose of the CNA Conference, or the frequently suggested surrogate of test or exercise data? Can it be expert judgment, the considered opinion of an experienced military officer as to what constitutes "reality?" Or can it be represented by the body of results forthcoming from a second model that is already, in some sense, "established?"

Each of these possibilities was suggested at the CNA Conference as having some advantages, but was also found to have its problems. Combat data, though they have the virtue of coming from a real war, are notoriously difficult to obtain, and they are often very inaccurate. (Chapter 1 on wartime combat modeling gives some examples.) In tests and exercises one is often able to measure more exactly than in combat, but what he measures is circumscribed by the rules and resources of the test. (Chapter 11 on tests and exercises discusses these difficulties.) Some of the conferees suggested that expert judgment, especially when one has a consensus of experts, can be a valuable source of insight, but others expressed grave distrust, pointing out cases where agreement of model results with a consensus of military experts might indicate that the modelers had succumbed to "client pressure." (Helmer and Rescher (1959) give a very thoughtful discussion of problems involved in using expert knowledge.) The use of other models as standards poses some of the same problems as the use of tests and exercises and the use of expert judgment, while it does "inherit" some of their advantages.

The basic point about "reality" is the non-existence of general agreement as to what it is. The approaches to it, suggested above, are incomplete and indirect. As in science generally, the dichotomy of "fact" and "theory" is an oversimplification. What is perceived as fact is likely to involve habits or patterns of perception that themselves constitute a sort of "model." In effect, we may be modeling our own mental "models" more often than we realize, and while this is by no means all bad, it is far from being as simple and clear a situation as the "fact versus theory" characterization would suggest.

Moving on from the difficulties of ascertaining "reality," it is important to note that even if this basic problem were solved, significant obstacles to verification would still remain. One may illustrate these obstacles with three kinds of difficulty, identified somewhat "poetically" as:

> "The Teleological Problem" -- How a model by its nature formulates an explicit cause-and-effect relationship that excludes other proximate or romote causes.
>
> "The Epistemological Problem" -- How the "truth" of any model is always provisional and dubious.
>
> "The Uncertainty Principle" -- How the very act of formulating or exercising a model distorts the reality we seek to represent.

In strict usage, "teleology" pertains to *final* causes, while model limitations constrain the representation of other causes as well. The terms "teleology" and "teleological" will be used here, however, to symbolize all of the model limitations that relate to the difficulties of representing cause-and-effect relations in models. These "teleological" difficulties insure a certain kind of inherent non-verifiability in any model. More generally, the "epistemological" problem insures that verification of any model must remain imperfect; this is, in effect, the central problem of the philosophy of science. Finally, the "uncertainty" principle of modeling combat has the effect of undermining in application some verification already thought to be accomplished. Therefore, these three serious difficulties, capsulized by the terms "teleological," "epistemological," and "uncertainty," all deserve further discussion.

Wayne Hughes eloquently introduces the "teleological" problem:

> Teleology is the study of final causes. A model always asserts a certain cause and effect, even when it has sophisticated feedback loops.... We presume a cause when we write inputs.... The model not merely asserts presumed first cause, but circumscribes for its user the world of admissible causes.
>
> Consider a warfare example: Why did Lee lose at Gettysburg? Historians may take as proximate cause the ill-conceived charge of Pickett on the third day. Or possibly Meade's artillery, massed in the center.
>
> As causes "once removed," there was Meade's astute tactical leadership and Lee's uncharacteristic tactical error. But few historians stop there. The cause was "really" J.E.B. Stuart's absence, so that Lee fought blind. Or the earlier death of his stalwart Stonewall Jackson.
>
> Deeper still, it was simply the inevitability that sooner or later the odds would catch up with Lee, and his daring battlefield tactics would overextend him. The fundamental cause, therefore, was the Union's greater mobilization base. Lee was impelled by a sense of urgency, knowing that time was against him. Thus, what historians may call a tactical blunder was Lee's last-gasp gamble, a gamble made with a thoroughgoing appreciation of the true odds against breaking through the center.
>
> None of the "causes" above is unimportant, and the list is by no means exhaustive. One could add the Union quartermasters' efficiency ("logistics dominate war"), the motivating reasons why the soldiers fought tenaciously, etc.
>
> All the "causes" contributed to the effect: the Confederates lost the battle. Any model of it will emphasize some things and deemphasize

> others, even to the point of complete exclusion. Whether the model is the analyst's simulation or the historian's description, it circumscribes the event with some set of cause-and-effect relationships. Any model, even the most ambitious, is vulnerable on grounds of sufficiency -- its omission of the n-th order "cause-of-a-cause-of-a-cause...."

When one contemplates Hughes' very interesting military example, he may be reminded of Goedel's incompletability theorem for mathematical systems: in a system at one level it may be possible to pose a problem that defies resolution at that level. It is necessary to go to a higher-level system. Here, the "teleological" problem of models gives us incompletability with a vengeance: there are some problems whose very statement requires a "higher-level" model. For example, in one of the "piston" (or single corridor) models of land warfare, as strictly interpreted, there is no representation of flanking maneuvers. Non-verification of the "treatment" of flanking maneuvers in such a model is "built-in."

In actual usage, of course, models are typically applied to situations not strictly represented. This may be done deliberately, and one then speaks of ingeniously "gimmicking" the model. If, for example, a land warfare model does not strictly represent supporting air power, but does represent artillery, then in application air power may be represented as artillery. Strictly speaking, however, any verification of such a model in such an extended application should include verification of the "rules" of extension. How, that is, does one represent air power as artillery? How are the numerical values of the inputs to be selected?

Careful consideration of the example above suggests the kind of problem we may have when we don't realize the "teleological" limitations of a model. The causes that are represented in the model must then serve as surrogates for those that are not, and we may on occasion not be aware of the model's deficiencies. Consider, for example, the "stopping rules" used in land warfare models. Often these take the form of casualty thresholds: an advancing unit may come to a halt, for example, when it has sustained x% casualties. Such rules were questioned at a conference a few years ago when a retired officer recalled that units under his command had halted when communications (and consequently control) had deteriorated to some intolerable level, rather than when force levels had reached a critical threshold. In some situations, of course, the decline of communications is highly correlated with the decline in force levels, and it would be difficult to disentangle their separate roles. One might, therefore, think that a model was verified only to discover later that such "verification" was limited in scope by the "teleological" limitations of the model.

Even within a recognizably limited scope, the verification of a model is also beset with a host of what may be termed problems of "epistemology." Epistemology deals with knowledge -- and in particular, with how we learn and what we learn from experiment. We encounter epistemological problems as soon as we probe the meaning of the simple recipe to test a model by asking if its results are realistic.

What the recipe says in effect is to test model M in a given situation by determining model results R, and then comparing R with "real" results, say R', which come from combat data or tests. In logical format, we have essentially:

If M, then R.

Observe R'.

R = R' ?

Conclusion = ?

The conclusion would be phrased in terms of verification (as, e.g., the model fails its verification test for the given application, or the model has passed another test and is closer to verification, etc.). It will be useful to use the format above in a discussion of the epistemological problem.

It has already been noted that the determination of R' depends on adopting a standard or source of "reality." This difficulty has already been discussed. It remains a serious problem.

Going to the next step, one asks: given both R and R', are they equal or nearly equal? This question, which seems very simple in some contexts, raises a multitude of difficulties. There are, first of all, measurement problems. R' is likely not to be known exactly. Combat data are obtained under difficult circumstances. Range instrumentation is imperfect in some cases, and test results are not always precise. Expert judgment may not lend itself to precise, testable pronouncements.

Such difficulties intensify when R and R' both come from probability distributions. One is comparing a sample of model results with a sample of results from the "real world." The question is whether the two samples may reasonably be considered to have come from the same distribution. When enough is known about the probability distributions involved, this may become a rather standard statistical problem, but the situation is often not that simple. It is especially likely that knowledge of the R' distribution will be highly imperfect, and such gaps may call for a great deal of judgment.

In any event, the program of verification requires a judgment as to whether R = R' -- at least to a reasonable approximation -- or whether it appears that R ≠ R', even allowing for measurement and sampling errors. Which of those judgments is made then affects the conclusion one would draw.

Consider first the case where one observes an R' in good agreement with R. What may he conclude? In a strict logical sense, one cannot conclude that the model is verified. He would surely feel, however, that the model is now more "plausible." Polya has endeavored to create a "calculus of plausibility" that would apply to such problems as verification, and it was suggested at the CNA Conference that considerations of plausibility may lend themselves to statistical decision theory. Such efforts would still appear to be in their infancy. More important is the observation that however one calculates it, what he has from one more instance of agreement is increased plausibility and not absolute verification.

The impossibility of absolute verification is perhaps most obvious when one thinks of using a model to represent a conflict that has not yet occurred and is of a kind not yet experienced. Such a conflict might well be expected to yield results R" diverging sharply from any real results R' previously used to verify the model. The differences between R" and R' may involve changes in time, in weapon mix, in climate, in terrain, or in tactics. Thus, one sees that the future conflict may well lie outside the domain considered in previous efforts to verify the model.

As the Overview shows (page 5), there is a range of applicability for almost every operations research model. When models are used for making predictions, it is especially clear that the range of applicability must be conjectural.

What of the other case, where R' definitely differs from R by more than the tolerable margin of error? Does one then discard the model? The answer, based on general scientific practice, is "by no means." The model M has failed a verification test, but what is "M"? Typically M is the conjunction of some P and Q, where P represents the model "structure" *per se* -- a set of postulated relationships -- and Q represents the inputs -- quantitative data. A deviation of R' from R casts doubt on the compound M, but the culprit could be Q (the data) and not P (the basic model structure).

Even, however, if the culprit turns out to be P itself, it would be premature to reject P totally. R and R' are likely themselves to be compound propositions. While P may not be adequate to model the situation to which R and R' correspond, P may be a good model for a more restricted problem. By restricting the range of applicability advertised for the model structure P, one may still salvage something from previous verification attempts.

Nor is this the last hope for P. Writing of theories and good scientific practice, Feyerabend (1975) observes that "*no single theory ever agrees with all the known facts in its domain*." He provides as evidence:

> Newton's theory of gravitation was beset, from the very beginning, by difficulties serious enough to provide material for refutation. Even today and in the non-relativistic domain there "exist numerous discrepancies between observation and theory." Bohr's atomic model was introduced and retained in the face of precise and unshakable contrary evidence. The special theory of relativity was retained despite Kaufmann's unambiguous experimental results of 1906, and despite D.C. Miller's refutation (I am speaking of a refutation because the experiment was, from the point of view of contemporary evidence, at least

> as well-performed as were the earlier experiments of Michelson and Morley). The general theory of relativity, though surprisingly successful in some domains,...failed to explain 10" in the movement of the nodes of Venus and more than 5" in the movement of the nodes of Mars....

If one is to begin to understand why such models are accepted or rejected, he must take a far broader view, as pointed out by Dr. Wilbur Payne (in correspondence with the author). Payne noted the inadequacy of looking at Bohr's atomic model in isolation, and solely in terms of the precision with which it predicted the frequency of radiation by hydrogen. The Rydberg empirical law had earlier given the same precision. What Bohr's model did was, first, to allow the Rydberg constant to be calculated from more fundamental constants and, second, to do this by using the idea that the energy of light is quantized in proportion to its frequency. Significantly, this was the idea that Planck introduced to explain the blackbody spectrum, and the idea that Einstein used later in explaining the anomalies in the energy of photoelectrons. The impact of this idea on physicists of that day would, as Payne points out, be explained by the concept that "a theory that agreed with experiments to about one part in a thousand in three experimental fields as widely disparate as these must be essentially correct even if incomplete."

One sees from such examples that the epistemological difficulties are severe. There is not a simple, mechanical procedure either for accepting a model as verified or for rejecting it because of disagreement with "reality." In either case, other properties of the model must be considered -- and considered not in isolation but in the broad context of scientific thought of the day.

Finally, in addition to the "teleological" problem and the "epistemological" problem, there is an "uncertainty principle" that plagues model verification. The presence of an observer on the battlefield, seeking to collect data, may have some effect on the combat itself. Even without an observer actually present, there may be distortion (as, for example, in World War II aerial bombing when some crews refused to drop bombs in certain unfavorable conditions after bomb cameras were installed in their planes because the combat film was used in a scoring system associated with efforts to improve the modeling of bombing accuracy). Thus, the act of modeling itself may alter the "reality" it seeks to model.

Verification as Process

If one concludes from the previous recital of difficulties that there is no simple procedure insuring the verification of a model or family of models, then he would be in good agreement with most of those who have examined the problem closely. He can, for instance, sympathize with a concluding comment of the Chairman of the 1968 CNA Conference:

> Surely out of all this, somebody must have figured out some good snappy answers to the whole verification subject. I am going to read the proceedings carefully. It must have passed by somewhere.

This conferee *has* recently reread the proceedings carefully. Pondering on them, and on much that has happened in the fifteen years since the conference, one may conclude that the answer is not to be found in any single subsection of the proceedings, but in the proceedings in totality, in the act of holding that conference and other similar meetings, in the peer criticism thus engendered, in the dozens of review groups that have examined models and the modeling problem, and in the rules that have been conceived and tried and revised. In sum, it is the total process itself that is the key.

One analogy used by this conferee is that of seeking to live the good life -- to attain perfection of character. The goal is never reached, but the struggle is key. The search for model verification is similar. A model, or family of models, is never verified absolutely in any substantial domain. Verification is always provisional. The sheer dynamism of what we seek to model should make that evident, quite apart from epistemological difficulties.

Verification efforts, the many arduous endeavors to make our models better and more useful, are absolutely essential. The struggle for excellence must ever continue. The resources devoted to models and to their use in studies are significant in themselves, but they are dwarfed by the vastly greater resources devoted to the weapons and forces that provide the impetus for those models and studies. The critical impact of force structure decisions demands the best of our tools and the skill with which they are used.

We must realize, however, that our verification efforts will not result in absolutely verified models certified for all time. Model use will remain conditional, dependent on the decisions to be made and the resources available for study. The user of a model will continue to have the responsibility for its use. He must be prepared to explain why he did what he did, and to show its relevance to decisions to be made. Previous verification efforts, helpful as they may be, can never provide in themselves complete justification for the use of a model.

Given the characterization above of "verification as process," it is difficult to do justice to the role of guidelines and rules. On the one hand, we should recognize that study organizations, as parts of still larger organizations that deal with highly important problems and vast resources, must have rules. Moreover, such rules can be exceedingly helpful (as, for example, when a rule to test a model against relevant historical data leads to rejection or correction of what would have been a very bad model).

On the other hand, we must remember that no written rules can establish final standards of model adequacy or capture all wisdom on verification. We must recognize, therefore, the danger that rules can be misused (as, for example, in making too strong either an endorsement or a rejection of a proposed model). In a large bureaucracy, with important decisions depending on the use of a model, there will be the temptation to make unwarranted -- even outrageous -- claims of verification or non-verification. In peer review we should never let outrageous claims pass unchallenged; but neither should we demand unattainable perfection. Decisions must be made, and a model may be evaluated as very helpful to decision makers for some decisions without meriting a general certification as a model verified for all seasons.

To the author this is counsel of hope, not of despair. The struggle to improve our models is key, and verification efforts are essential. Though we can never attain perfection, we can do much. The struggle is worthwhile, and it should be an enlightened struggle. We should never let the unattainability of perfection lead us to seek less. We must do all we can to avoid both the sin of sloth and the sin of pride.

REFERENCES

1. *Center for Naval Analyses (CNA), CNA RC 115, Warfare Model Verification Conference, 1 October 1968.*

2. *P.M.S. Blackett, Studies of War, Hill and Wang, New York, 1962.*

3. *Paul K. Feyerabend, Against Method, Atlantic Highlands: Humanities Press, London, 1975.*

4. *Olaf Helmer and Nicholas Rescher, "On the Epistemology of the Inexact Sciences," Management Science, vol. 6, no. 1, pp. 25-52, 1959.*

5. *Philip M. Morse and George E. Kimball, Methods of Operations Research, OEG Report No. 54, Operations Evaluation Group, Office of the Chief of Naval Operations, Navy Department, Washington, DC, 1946.*

CLAYTON J. THOMAS

The biography of Clayton J. Thomas appears following Chapter 1.

ABOUT CHAPTER 14

In this chapter Wilbur Payne discusses the problem of model validity in both technical and management terms. As a technical matter he notes that models are, by definition and necessary practice, invalid in at least the sense that they contain abstraction, the omission of observable details, and approximation in computational approach. Beyond that he points out that models can be intrinsically invalid in their logical structure and, even if they pass that test, invalid in application because of the input to them. He discusses some of the limits of history and experiment as the ultimate test of validity for predictive models.

As he says, "Models may be right, but misapplied." Granting the necessity to seek resolution of questions of model validity in experiment and history, he argues that even these must be supplemented with careful and competent review processes. The keys to this, in his view, are an open planning process that makes review possible and the existence of a family of models with different abstractions and approximations.

We see then why an analyst is disconcerted when asked whether he has validated his model. The answer must always be no. The correct questions are: What steps has he taken to corroborate his model, and over what range of applications can it be shown to be utilitarian? The reader of this and the preceding chapter is unlikely ever again to ask the wrong questions, but will with renewed confidence let a good model be put to good use for the purpose it was intended.

Validity of Models as a Basis for Military Planning

by

Wilbur B. Payne

The general question of validity is one management problem inherent in the use of models as an aid to military planning. This question would exist however simple or complicated the models employed. But it is much more difficult to answer when asked about results derived from a large family of interacting models, each of which more or less fills to capacity the largest computer available to the agency which designed and uses the model. It is clear from existing reviews of models and of a model-based planning process that much confusion will be created unless it is realized from the outset that questions of validity are not simple.

In one sense all models are invalid. They are, after all, abstractions. There always are observable properties of the system being modeled that are not represented in the model explicitly or implicitly.

In addition to the fact of abstraction, models characteristically exhibit approximation. For example, processes known to be non-linear will be approximated with linear relations. There will thus be an assumption, explicit or implicit, that no application of the model will get outside of some range of variables within which its approximations have adequate precision.

Abstraction involves the omission of some observable factors or interactions between them because of a belief that they do not have important effects. Approximation involves limits on the precision of calculation of the effects of the factors or interactions that are present. Common approximations are the rounding of numbers to produce what are called "significant figures," computing a function with the first few terms of a converging infinite series, and the aforementioned approximation of non-linear processes linearly.

There is, in the context of predictive models, a connection between abstraction and approximation. That may best be seen through standard textbook treatments of theory of errors. Even the best experiments to measure the present state of a system have limits in precision. The limit of precision or a band of uncertainty associated with prediction of the future state of a system increases with both the time horizon of the forecast and with the inherent complexity of the system. In one sense, then, abstraction is possible (as well as necessary) because one knows *a priori* that some factors will have effects small compared to the intrinsic uncertainty about processes or to the lack of precision in necessarily approximate computation.

It is possible through review of the structure of a single model to establish that it is invalid because it contains a logical or mathematical error. But if a model passes that test, then any further pursuit toward the question of model validity always involves appeal to some other model or models and to the context of model applications.

At first that sounds strange. Many writers on the subject have noted that in other disciplines that make extensive use of predictive models there is an ultimate appeal to observation and experiment. This is also true for the family of models used in military planning. The similarity between the use of observation and experiment in military operations research and their use in science or engineering is at least as important as any difference. To illustrate this, note the following:

1. Observation, both as military history and in the form of peacetime field operations, is employed both in development and review of models.

2. Most modeling agencies in military operations research have access to experimental facilities. The Army, for example, has two large agencies for the conduct of field experiments whose purposes are primarily operational.

3. Experiments of a scale requiring the formation of a division force and operating it in the field for several months have been undertaken three times.

The models employed in military planning have some intrinsically untestable predictions. They deal, among other things, with weapon systems for which there is not yet a working prototype. But many of the models employed in economics and astronomy have this property, as do some employed in engineering. There are, from time to time, predictions from military models or important inputs to them that could be but have not been tested. But anyone familiar with physics or engineering is aware of predictions that could have been tested but were not. There are predictions from the models of physics or engineering, as from those of military operations research, that cannot be tested when the predictions are first available because experimental methods cannot yet test them -- as, for instance, the prediction of the existence of the planet Pluto or the prediction of the gravitational effect on the path of light. Even more frequently there are predictions and input parameters that are not tested because of generalized management judgment that the costs of empirical validation outweigh the benefits from such activities.

The next step necessary to the argument that validity of models rests on appeal to other models is to realize that both experiment and observation are "models." The discussion of field experiments in this monograph makes this clear for experiments. Field experiments differ from mathematical models only in which aspects of the system they choose to treat explicitly. But, clearly, they do not and cannot avoid abstraction and approximation and hence are "models."

For history as a form of observation it is necessary only to note that we have access to reports of what occurred in the past but not directly to the events. But the same is true of today's reports about yesterday's field exercises. As with mathematical models, written history or even the memory of participants contains demonstrable abstraction. A judgment of importance or relevance always underlies the decision of what facts or factors are to be omitted from the report or to be forgotten.

As with other scientific disciplines, conflict between mathematical model predictions and history, observation, or experiment are matters of import. It is clear from other sciences that there can be faults in (as well as false) observation and experiment. Hence, there is no historic basis for a belief that every conflict between model predictions and an existing observation or historic report can be resolved by immediate acceptance of the observation or experimental results and rejection of existing mathematical models.

Finally, anyone who reads must be aware of directly conflicting results -- even between experiments -- as well as conflict between experiment and mathematical models. Nor is it uncommon to find disagreement among historians in their description of the same events. Thus, it is clear that every model, including history and experiment as models, is in some sense invalid and that, however valuable they may be, observation and experiment are not infallible determinants of validity.

There is a further complication. Nothing but confusion results unless it is recognized that the question of model validity comes in two different flavors. Asked by a manager who might use the output of a study as a basis for a program or budget decision, a form of the question might be: "How do (or can) I know that model output is right?" The same form of the question applies to those who want to use the results from one cycle of the planning process as input to their own part of it. From this point of view, an answer to the question must consider not only the logical validity of the model structure and the precision of the computations within it, but the validity and relevance of the planning assumptions which are the input to the model.

Embedded in this broader question is the narrower one that can be applied to a model as a set of instructions on how to do certain calculations given certain inputs. From this viewpoint, the correct form of the question is: "How could I know if the model is wrong?"

The form of the question of validity asked by managers is the more general of the two. It must encompass both the general planning assumptions (called input data in modeling jargon) as well as the validity of the model structure. The form of the second, narrower question asked of or by the model designer is the most that can be answered through inspection or review of a particular model. The planning assumptions that will be provided as input to a particular application of the model will vary from time to time and may not even be known or suspected at the time a new model is tested and reviewed for validity. In fact, some planning assumptions that are current when a new model is reviewed originate at such high echelons that they are called policy. They are not likely to be challenged as part of a model review process.

An historical allegory might help. In the period beween the two World Wars, France adopted a defensive concept and forces and systems to implement it. They had, of course, just finished the first war in modern history in which the principal and decisive operations were defense of and attacks upon a flankless, linear defense. The defense was not really flankless in World War I, but neither force could exploit the flanks -- one of which was the Alps and the other the English Channel. Assume that instead of the manual war games and field exercises available in that time frame, the present family of ground combat models had been available to help in the design of the system and forces to implement this concept.

It would be natural now, as then, to ask some questions about the threat. "How and with what will the Germans attack the Maginot Line?" would seem natural. With that, one could turn to a dynamic force-on-force combat model representing some significant segment of the Maginot Line and explore both detailed and general questions of system and force design. In repeated and iterative use of this model it would naturally get sharper and more refined. Factors put in initially as crude estimates of weapon performance and operational factors would, as actually happened, turn out to be important, and more precise estimates would be generated. There was, for example, an extensive body of literature developed analytically and empirically before and during World War II on the number of hits and the number of shots by artillery of various calibers required to damage reinforced concrete pillboxes of various designs.

What would be different in 1982 from 1932? Today's models would tackle this planning problem in bigger bites. There would be a somewhat less fragmented planning process and perhaps fewer steps of iteration. Each of these steps could pursue the problem with a great depth of technical and operational detail. On the other hand, today's models would probably report, as did yesterday's, that the Germans could not penetrate the defensive system in any reasonable time.

Suppose at the end of the process one had developed a model that took this problem one corps sector at a time and that the model passed whatever tests or checks of validity anyone can imagine. *It could not have helped if applied as described above.* As described above, the model was employed to test a *system* of defense. The German attacks, when they came, tested the *concept* of the defense and found it deficient. A corps-level model is flankless. The Maginot Line, as events developed, was not.

The lesson from this may rest on the observation that it was necessary but not sufficient for French military planners to test the *system* of defense. Recognition of this necessity led to creation and application of models for that purpose. Failure to recognize that this would not suffice led to failure to develop or exercise models that could test the *concept* or challenge the planning assumptions.

Naturally, no model can help answer an unasked question. Nor is it clear that the French had no models that could have tested the *concept*. Any models they actually used could only have been descendants of Kriegspiel. Kriegspiel was developed in an era when wars of maneuver and open-field battles were common. Even if some descendants of Kriegspiel became specialized toward the problem of defense of a fortified zone, the earlier versions did not cease to exist. They simply ceased to be used.

Suppose, to continue our allegory, some of the opponents of the concept of defense that led to the Maginot Line had employed an open-field version of Kriegspiel to lend support to their opposition. Would the model have been found valid? Almost certainly not. With years of disuse it had probably rusted. It would not have contained up-to-date representations of either French or German forces. Given that the responsible agencies and senior officials were all proponents of the existing defense program, the fact that the model was probably good enough to raise serious questions about some long-cherished planning assumptions could have been hidden in a long and technically correct list of model defects and deficiencies. From this it is reasonable to observe that it is very unlikely that a "valid" model will exist the first time a question is asked.

How can a manager approach the general problem of validity of models and of the planning process they support? If he agreed with the preceding discussion, some necessary elements of his course of action would be clear. He would, for example, see the absolute necessity of an open planning process and for a substantial program of research in parallel with applications of models to specific problems. In spite of the general requirements of security, he should recognize that none of the three elements of his problem -- planning assumptions, model abstractions, and approximations -- can be challenged unless they can be exposed to review. He would see that models can be wrong, or they can be "right" but misapplied. He would see that issues of validity or invalidity of either model structure or application are resolved, if at all, in context of the larger family of models that will include experiment, history, and observation of force operations.

Our hypothetical manager probably cannot afford the *laissez-faire* policy of "Let a thousand flowers bloom." But he certainly should not reduce his nursery to a single hothouse orchid. Our manager probably cannot personally review the narrow technical validity of many of the models in the family, nor each and every planning assumption employed in their applications. But he can and should develop a system and foster a climate within which such reviews are routinely conducted. And when such reviews lead to a suspicion of errors with important consequences, he should have in place adequate research and experimental resources to resolve the issues. The general character of this necessary review process is clear. It rests on the interaction between inquiring minds and various members of the family of existing models. In the best sense of the word it is thus "scientific." Some of the specific characteristics of the necessary review can be illustrated by discussing some aspects of the current family of models.

Mathematical models have, as one of their more important properties, the ability to demonstrate the consequences of assumed error in any of their inputs or in their mathematical approximations. Thus, they can give some usable information about which of their always uncertain inputs should be made less uncertain if possible. Thus, one aspect of the review process should be the exercise of existing models to exploit this property. This process is generally called sensitivity analysis although, more often than not, it is limited in practice to testing sensitivity to input factors.

Important as it is, sensitivity analysis limited to exercise of a single model is not sufficient. It cannot tell one anything about the consequences of facts or factors that were eliminated as part of the abstraction from reality inherent in the model design. To deal with the question of omissions in a particular model it is, in fact, necessary to appeal to other models within the family of models if any can be found which include the effects of the omitted factors. The fact of omission is always present since models are abstractions. Contrariwise, the fact of omission cannot automatically be said to make a particular application of the model invalid.

To illustrate this, there exists within the model family some models designed to help estimate how many of what kind of repair parts should be stocked to sustain operations of the helicopter fleet within a division force. The model is, in every reasonable sense, a land combat model because it must represent the activities of the helicopter fleet and, among other things, the damage that might be inflicted by hostile fire. (Hostile fire can, of course, generate demands for repair parts for the helicopters that are hit but survive, and supplies of spare parts from helicopters that are hit and do not survive but which do reach friendly-held terrain.) The model does not contain any representation of naval gunfire or of nuclear missiles. Such things exist in the hands of potential opponents, so the model is in some sense invalid. However, the model has been used for several years in parallel with other models at a higher echelon of the hierarchy. In all that period, the division forces involved have not conducted operations within 100 miles of a sea coast nor been attacked with nuclear weapons. Does this lack of "validity" imply a lack of "utility?"

The *family* of models should be declared invalid and would be more dangerous than useful if models within it could not represent all components of the threat and all tactical employments thereof. No such standard can be applied to each of the individual members of the family. Application of that standard to the family models is more reasonable. But even here the appropriate and valid models may not exist only because there has been no recent management interest in such problems or because certain general planning assumptions make development and use of such models appear to be unnecessary.

Another aspect of the review processes necessary to the validation of the family of models, or its individual members, is the recognition that no finding of validity through review is final or permanent. Particular models (or at least their acronyms) may have a long history of use, modification, and review. But all too frequently one finds within them features that were once justifiable omissions or approximations but are no longer acceptable. Technological progress or changes of organizational and tactical concepts can make once acceptable approximations dangerously inadequate. Nearly every application of a model of the sort used for military planning requires careful review and usually modifications to incorporate new information that has emerged from other parts of the military planning cycle including research, tests, and experiments conducted both inside and outside of the defense establishment.

Within the family of models there are models with quite different levels of abstraction and large differences in their precision of computation. That is a strength in the sense that, although many models accept linear approximation to non-linear effects, other models (and usually experimental techniques) are available to check the adequacy of such approximations. By their very nature, large-scale computer models can be exercised with wide variation in input parameters and, at somewhat higher cost, with variation in the precision with which certain activities and processes are represented. In this mode of operation they can develop information about the consequences of uncertainty about, or errors in, their input or level of approximation.

One of the obvious dangers of a family of models is that an undetected error can propagate through all levels of planning. Each aspect of model design can to some degree be cross-checked. Questions of the form "Do any of the observables omitted from this model have significant potential effects?" can be checked by appeal to history or to experiment or to comparative results from models that incorporate those observables. Similarly, the input variables and level of computational precision can be checked. *But the actual process should not be viewed as one which is automatically self-correcting.* As is true of experimental validation, broad management judgment intervenes and available comparative tests that could provide a deeper insight into validity of particular results are not always fully exercised.

It is possible and, in fact, frequent for a kind of bureaucratic mythology to develop such that many people and agencies believe that what they already know about the future is entirely adequate to their needs. One of the symptoms of such mythology is the existence of many studies -- all constrained to a very narrow range of planning assumptions. It is also characteristic of such periods that the complete range of safeguards, such as sensitivity analysis, available within a model-based planning system will not be employed.

The potential of models to identify the areas in which uncertainty about some aspects of the future have most important impact, their ability to record in detail the facts and assumptions on which earlier planning was conducted, and the ability to explore the consequences of omitted factors in one model through appeal to others are, when fully and objectively exploited, the important strengths of a model-based planning system. Their diligent use in an open, objective, planning environment was very probably one consideration at the time of the post-World War II decisions to create or expand the operations research activities in the military departments.

What was seen as fundamentally different in the system proposed from the system that existed before and during World War II was the incorporation of scientists and scientific methods in military planning activities. Scientists had long been involved in weapon design and development. Many more are employed for that purpose, even now, than in operations research. Events both during and especially at the end of World War II led to a realization that continuation of rapid change in weapons and other materiel systems was both necessary and inevitable. There was adequate evidence not unique to World War II, but certainly highlighted there, that considerations of how forces would be organized and would employ new weapons was deserving of as much rigorous investigation as how the weapons should be designed. In the thirty-odd years since these decisions were made, there has not been a controlled parallel exercise of the pre-World War II planning methods with the new ones. So objective comparison is hardly possible. On the other hand, there is also no reason to challenge the view of senior military officers who had been involved in prewar planning and in the planning and conduct of wartime operations. Their view was that pre-World War II planning methods could not cope with the emerging problems.

DR. WILBUR B. PAYNE

The biography of Dr. Wilbur B. Payne appears following Chapter 4.

ABOUT CHAPTER 15

In introducing the two previous chapters, we showed the importance of asking the right questions about model validation and verification. Now, in closing, we devote the last chapter to asking right questions in general. As Stephen Leibholz says, it is time to serve the *user* of the modeling product by arming him with the means to manage and evaluate the analysis.

It will be seen that Leibholz deals with more than modeling, for one cannot ask the right questions without examining the larger context of the analysis. Now and then, he raises some issues of analytical technique that are not easy for the beginner to grasp; if so, the exposure should do more good than harm, and the questions remain as useful as ever.

Leibholz also projects the reader out of narrow models and analysis into considerations of motivations and of roles. To understand why this needs to be, one must read no further than the first two of his 20 questions: questions of explicit and implicit sponsorship and the effects of a study's terms of reference. Having pointed out the transition, we do not think most readers will have any difficulty adapting to it. We take our readers (some of them) to be amateurs in the use of models but professionals (all of them) in the art of making the world go 'round.

We trust these "20 Questions" are a fitting capstone to our treatise.

Twenty Questions
by
Stephen W. Leibholz

INTRODUCTION

This chapter is intended to serve two roles: first, it will act as a complement to the documented experience of this monograph, coupling the art of modeling to the *rude*, real world where models are commissioned, managed -- and perhaps utilized. This experience transcends the structure and specific application of the models themselves, and thus places emphasis on the management-related inferences and lessons which may be drawn. Second, and perhaps more important, it provides a different perspective on the issues of modeling as applied to military operations: the perspective of the Government, military, or industry manager who has commissioned the work or who is to become the user of the results of the exercise of that model, and who has been called upon to make some hard decisions on programs, policy, or tactics -- decisions typically based on a brew of logic, history, psychology... and of modeling results and insight.

The client and manager are easily the neglected parties in the cast of players, critics, partisans, and auditors who associate themselves with a given project of modeling; there are many works and much advice addressed to the analyst, but little to the persons who are heirs to the products of his labors -- who must utilize them and be judged by them. To partially redress this imbalance, this essay is addressed to the client and his surrogate -- the manager who is (presumably) cognizant of his client's needs and responsible for the project's compliance with them. Since the span of the manager's detailed inquiry into a modeling effort must, therefore, include the client's concern as a subset, we refer to both in the following paragraphs with the term *manager*. And shortly, we will point out and utilize the implied role of each as *mentor*.

The "bottom line" of this paper thus comprises a set of criteria for the manager -- whether civilian or military, whether or not he is a specialist in the art of modeling. His needs are to manage and utilize the products of a modeling effort for the purpose of aiding his decisions. The guidance takes the form of the necessary dialogue between the model-builder and the manager-user, resulting in assessments which must be made at various stages in the analysis (including the original challenge to the analyst) and before the results are accepted.

The title "Twenty Questions" was chosen to suggest an analogy with the parlor game in which an object is to be identified on the basis of binary (yes/no) answers to 20 questions. Since optimal play of this game will, at best, allow the questioner to distinguish among 2^{20} or 1,048,576 objects, good players must therefore rely not on logic alone, but also (again) on a brew of logic, history, psychology...and of modeling results and insight.

The game is therefore a paradigm of the classical problem of the decision maker: *to perform successfully in the absence of a closed-form solution to his problem*. And neither is the model a closed-form solution.

THE MANAGER'S PROBLEM

The nature of the military environment has fostered a special role for military modeling in the intellectual arsenal of the military decision maker. Wars are never fought twice, despite the admonitions of Leo Tolstoy; and when they are fought, the participants have more on their minds than the recording of data for future analysts. Moreover, those wars of utmost criticality to national security have never been fought. Therefore, there is for military operations and systems the special role of modeling as a surrogate for empirical data.

These realities of military modeling present the client and manager with a problem. They must recognize the difficulties which the model-maker has in expressing the findings, limitations, subtleties, and unresolved issues of his model -- whether these difficulties are those of semantics, perspective, inexperience, or salesmanship. They must perform and then utilize the results of a penetrating sampling, an incisive "spot" examination, and draw evaluations and insights therefrom. They do not have the time to delve deeply into the model and sort the inferential roots from the square roots.

The management problems relevant to modeling and to its application in the "real world" tend to transcend the subject matter and objectives of the model -- thus, the "orthogonal" nature of this paper with respect to other partitionings in this volume. We are dealing with management problems which have a great degree of concurrency. The models and applications considered in this paper are obviously prescriptive rather than descriptive; directly or indirectly, they are intended to be utilized to influence plans and decisions.

The technology or specific mathematical constructs to be used in solving a specific management problem are by no means foreordained, although historical antecedents, personal proclivities, and the dictates of fashion may suggest otherwise. It is simply not true that, for example, a force-structure problem requires a human war game to solve, or that Monte Carlo simulation is essential to the analysis of engagements. The selection of modeling technology is a valid management question and is briefly addressed here. Further, once it has been selected, the questions to be asked have a degree of variation resulting from the variations which they exhibit in the concerns, interpretations, limitations, and pitfalls which are inherent to each of them.

We will not distinguish between those managers who are the clients or users of model results and those who manage the model-makers. Their interests are highly parallel. To the degree that a manager can direct or influence the modeling exercise itself without actually micro-managing the effort (in which case, he enters into a conflict of interest), he can ensure usefulness of the product; in any case, however, he is a user of the results. We do not, however, propose to concentrate on the separable albeit more difficult problems of managing professional people.

What is the role of the manager in the modeling environment? Certainly not a role of resignation and acceptance (a typical syndrome triggered by the contemplation of a foot-high pile of printout) nor a role of microscopic direction in which the premises, sources, methods (and perforce the conclusions) are directed by the manager. His role ideally is one of guidance -- of doing what is necessary to assure that the analyst perceives and formulates the problems in terms of the plans, decisions, and action required of the manager, and one of evaluating the results in the same objective terms. His role is thus of a *mentor*, about which we have more to say at the conclusion of this paper.

There thus is a strong analogy between the hoary parlor game and the mentor's problem: getting the right answer with limited time and facts. Unlike the rules of the parlor game, however, the "rules of engagement" of military management -- whether of systems or operations -- allow for probing, deliberately multi-sided questions, and for follow-up; each of these elicits far more information -- communicated, implied, and (most important) inferred -- than even the most precisely formed binary question. Thus, the additional role of the mentor.

THE INDIVIDUAL QUESTIONS

We have divided the topic, admittedly with some malice aforethought, into 20 broad questions. These are grouped into three categories as follows:

Those dealing with model externals:

1. What are the premises of the analysis itself?

2. What do the terms of reference say?

3. Where did the model originate?

4. What are the objectives and levels of comprehensiveness?

5. What are the inputs and outputs?

6. What assumptions and variations on these assumptions bias the model?

7. To what extent has the model been verified and validated?

Those dealing with model precepts and concepts:

8. Are the hypotheses falsifiable?

9. Should gaming have (or not have) been used?

10. Should simulation (or analysis) have been used?

11. What are the trade-offs of the number of players in the model?

12. Hierarchical models: something for nothing?

13. What was left out?

14. To what extent is the model design-specific?

15. How are human elements treated?

Those dealing with model internals:

16. How does the analyst deal with uncertainties?

17. What are the effects of hidden dependencies?

18. Was the model used within its range of legitimate applicability?

19. What are the effects of the underlying mathematical assumptions?

20. How many iterations are needed?

By no means is this list exhaustive, nor does it offer a road to instant expertise. Instead it is intended to offer some points of entry -- some opportunities to phrase incisive questions -- from whose answers our mentor may infer the degree of confidence which he is willing to place in the work.

A. QUESTIONS DEALING WITH MODEL EXTERNALS

1. What are the premises of the analysis itself?

Overriding all questions of form or substance is the honest and probing question of implicit sponsorship. An analyst may properly play the role of advocate or arbiter. In fact, in some sophisticated views, it is inescapable that an analyst is simultaneously an advocate of decisions to be made at higher levels and an arbiter, deliberately or inadvertently, of decisions offered at lower levels. This is complicated by the possibility -- which we explore later -- that the model may not be currently utilized at the level for which it was designed. Thus, we have the issue of the *role-relationship* of the analyst -- and of the broad expanse of grayness of intent -- which may indeed be no more within the "window" of professional integrity than the product of enthusiastic advocacy.

There is no simple diagnostic tool for this problem, which is somewhat ubiquitous in the profession, simply because analysts are retained for both judgmental and advocacy roles, and are restrained from admixing the two primarily by their personal standards of professional objectivity -- and by their good taste.

2. What do the terms of reference say?

MG Emmet Paige said it well: "Let me write the work statement and I can control the output of any study." This is an assertion of the power of subtlety, not venality. You will never see a work statement that says "Make certain that the answer is XXX." Instead, for example, the terms of reference might rule out critical but unwelcome factors, such as maintenance or crew costs, or might cite an unreasonable enemy capability. It is important therefore to determine if the analyst had the guidance, latitude, and resources to answer the manager's needs.

There is a second reason for examining the terms of reference: Did the analyst fulfill them? If not, why not? Perhaps the exigencies of the work itself demanded some modification of the original direction of work -- in which case, these circumstances must be brought out.

In summary: What were the *a priori* and *a posteriori* challenges to the analyst, did he fulfill them, and do they respond to your needs?

3. Where did the model originate?

It has been explained that a model may be *ad hoc* (that is, developed for the occasion) or it may be a *standing* model, requiring some degree of adaptation to the problem at hand. The former has the potential advantage of closer relevance. The latter may have a longer history of development and testing. In every modeling task in which this author has been involved, the model was not purely of one type or the other. Every *ad hoc* model has antecedents, and every standing model requires adaptation. The antecedents of the former may be highly removed from the present application and the adaptation of the latter restricted to a shaping of the data inputs, but all practical models have some aspects of both history and recent creativity.

An inquiry into the prior art is important because: (a) it lays open the premises which must be examined; (b) it suggests the constraints and limitations of application which may hold over from the prior use; (c) it allows the issues of model validation to be clarified; and (d) it may alert the manager to use of an inappropriate model to a present problem.

The interrogator should be particularly alert for signs of inappropriate model selection: dusting off and shining up a model long in the tooth but dear to the heart of the analyst or (more frequently) rejection of a perfectly applicable standing tool in favor of creativity and intellectual enterprise. The latter is more pernicious because the model that hasn't been tested is obviously free of all *known* bugs.

4. What are the objectives and levels of comprehensiveness?

A model can be designed for many breadths of scope. It may consider operations at the individual, squad, or platoon level all the way up to entire nations -- or it may consider systems from the micro-level of integrated-circuit structures to entire force levels. The scope of the model design must be ascertained. If the model is too general, it is likely to assume as "givens" -- perhaps implicitly -- precisely those variables of concern. If it is too detailed, it will require inputs of irrelevance to the present need, each of which must be derived from data or assumed -- most likely, the latter.

A recent attempt at mitigating this problem has involved the use of "nested" or hierarchical models, each at successively different levels of generality. This procedure, while useful itself, introduces alternative problems (see Question 11).

5. What are the inputs and outputs?

Here is one area in which detailed sampling by the manager is required. The model-maker is always on the horns of a dilemma: suppose the number of independent (or *input*) variables is large; as a result, the analyst gains flexibility, assurance that all bases have probably been covered, and a shimmering aura of realism. He also gains an intractable problem: that of *filling* these input variables with meaningful numbers. If he does not have data, he often uses random variables and falls neatly into a trap: by randomizing an unknown input datum, he effectively makes some very strong statements about its nature. (See Questions 6 and 17.) On the other hand, suppose the number of independent variables is quite small; he then runs the risk of suppressing an important parameter.

With regard to output variables, similar considerations apply. In addition, given a statistical variation of inputs, the output statistics demand a surprisingly large number of runs for any given degree of certainty of outcome. This number grows rapidly with the number of output variables. (See Question 20.)

In summary, check the input and output variables. Do not be swayed by the assumption that complexity equals realism or that complexity equals relevance.

6. What assumptions and variations on these assumptions bias the model?

The overt assumptions of a given model (that is, the stated ones) are, of course, the first-discovered sources of any biases which the model might exhibit. The wise modeler tests these assumptions by varying them to determine, at minimum, the sensitivity of the results to the assumptions; this provides the user with a means of assessing a qualitative "confidence interval." The existence and results of such sensitivity analyses should be checked. (Also see Question 13.)

More pernicious are the unstated assumptions which most often come about as analysts' oversights or misinterpretations -- or as programmers' fancies. Perforce they are never tested. Here is where the questioner's skill comes into play. We suggest you focus on one or two important variables which are difficult to quantify and which are either not stated or dismissed with phrases such as "taken into account" -- and that you require a tracing of their treatment.

7. To what extent has the model been verified and validated?

The recent rise of the phrase "Verification and Validation" as the name of a specific set of procedures for auditing software (and of models) has provided many advantages of use to the manager: among them, heightened awareness in a broad community of the need for and importance of such review; and the development of certain formal means intended to help assure completeness. However, there is a concomitant danger that the form might conceal a lack of substance.

"Verification and Validation" refer to two processes:

(A) The process of determining self-consistency -- that the software or other implementation of a model accurately reflects its description or specification. Even given an exhaustive line-by-line analysis -- which is often not economically feasible -- this can be better determined by running the model for a number of cases (particularly pathological or extreme cases) for which the outcome can be calculated. We refer to this process as *verification*.†

(B) The process of determining that the model reflects the expected application, the real world or requirements, and operational data. The success of this process hinges primarily on: (1) the treatment of overt and hidden assumptions, and (2) the level of detail or of accumulation which may or may not reflect the need. One way to sample-test the model is to run known historical or "reasonable" cases and examine the variance between expected and realized results. We refer to this process as *validation*.

A computer program is never a model. It is merely a computation and data manipulation procedure. Therefore the task of verification and validation must be applied to both the model and its computer program (if any). Ideally verification and validation are carried out independently of model construction, thus giving rise to the currently

---

†*Which of these processes refers to "verification" and which to "validation?" As often occurs when jargon overtakes English, there are two schools of thought on this subject. However, most DoD references cite Process A as verification and Process B as validation -- and so shall we.*

popular set of initials "IV&V" to denote a formal methodology for reporting the results of Independent Verification and Validation. Although applied primarily to computer software, the concepts are equally applicable to analytical models, regardless of the degree to which they are exercised in computer software.

Perhaps the best means of validation involves utilizing exercise data drawn from maneuvers or war games (themselves models) as inputs to the model, then comparing the outputs of the model with the results of the exercise. It is rarely possible to perform validation using combat data because the data are most often too sparse on the "input" side of the equation. Second to the use of exercise data is the use of "reasonable" cases to test the model. This approach is fraught with danger because it carries a predisposition to beg many subtle questions. In fact, it would be surprising if any surprising results emerged from a "reasonable" case test.

If the manager is faced with a lack of exercise data to validate the model adequately, he should then look for historical proof of the model's antecedents (see Question 3). Failing that, he can only direct the validation of the model "in the small," using data which test portions of the model one at a time; then somehow take into account the interactions and scaling effects which may never be tested. An example of the latter necessity occurs in many-on-many gaming simulations and in simulations involving models of human behavior. For these two classes of models, experimental data are difficult to acquire in any quantity (primarily because of the cost); these models most frequently have to be tested "in the small" and subjected to a diagnostic analysis to assess the interactions and scaling effects among the model elements.†

As a practical matter, validation of consistency with model specifications can be accomplished fairly thoroughly by verification against real-world data only to some

---

†*Indeed the difficulties leading to testing "in the small" often apply to expensive weapon systems themselves, particularly to hardware that has never been used in anger. What makes this limitation dangerous is forgetting the fact.*

degree of reasonableness -- and then with great difficulty. The manager will almost certainly be forced to limit his confidence in the results of the analysis by the degree of testing which had been achieved, and therefore the assessment of how well IV&V can be (and was) conducted is an important part of his duty.

B. QUESTIONS DEALING WITH MODEL PRECEPTS AND CONCEPTS

8. Are the hypotheses falsifiable?

Models are used to test hypotheses. Sometimes they are used to assign values to dependent variables, which, in a sense, constitutes testing continuous sets of hypotheses. The question, then, is one of determining whether a model tests the hypotheses it purports to test. These hypotheses must be stated explicitly. Are they? Now the basic scientific concept of *falsifiability* comes into play. Only if: (1) the negation or complement of each of the hypotheses makes sense, and if (2) in each case both the hypothesis and negative alternative are *feasible outcomes* of running the model -- can one accept the assertion that the model tests the hypotheses.

> Example: (1) "Tactic A is superior to Tactic B in terms of exchange ratios."
>
> Falsification: (2) "Tactic B is superior to Tactic A in terms of exchange ratios."
>
> Question: Can a reasonable run of the model yield (2)?
>
> An example of an hypothesis whose falsification demonstrates its weakness: (1) "Tactic A is always superior to any other tactic."
>
> Falsification: (2) "There is at least one tactic which under one condition is superior to A."
>
> The hypothesis is therefore silly because (2) is trivially true.

The test of falsifiability is a powerful tool for the evaluator. Does its falsification yield a reasonable alternative hypothesis? If so, can a demonstration of the original and falsified hypotheses each be a feasible outcome of exercising the model?

9. Should gaming have (or not have) been used?

Gaming, in the broadest context, is modeling which incorporates an adversary; it is a way of introducing the effects of competing human minds into an analysis, whether the human elements are represented by people or by algorithms. In the former case, gaming includes, for example, war games (see Question 15). In the latter, it includes techniques such as game theory, which makes some strong and questionable assumptions about human behavior, as well as more plausible models. Gaming is frequently either unrealistic or expensive, time-consuming, difficult to replicate and verify -- and hence no unmixed blessing. The user needs to observe the disadvantages and kind of biasing which might occur. Does the investment in gaming excessively reduce the investment in other aspects of the model (e.g., the variation of parameters)? Does the model account for the varying personal styles and risk aversion characteristics of the players? How were the choices of the players constrained? What is the effect of the relative lack of repeatability of the model if human players were used? Are the players representative of the population for which they are surrogates? To what extent do the gaming aspects of the model mask the cause-effect relationships among the input and output decision variables? (Also see Question 15.)

10. Should simulation (or analysis) have been used?

This problem has been discussed at length from a theoretical point of view, so we will review some pragmatic matters that often supervene. A well-executed analytical model can provide answers in less than prodigal time -- answers both *as sound and as defensible as any simulation*. It is, however, true that an analytical model can easily be made simplistic -- most often by force-fitting the reality to an academic framework for which both the software and the jargon are well-developed.

Each class of analytical tools has its own questions based on the mathematical assumptions which underlie them. If it is a linear programming or game theory model, is there a basis for assuming that the objective functions and constraints are linear? Does X knots of speed always equate to Y nm of range? If it is a Markovian or a Poisson queueing model: Are the statistics of waiting time in each state exponential? (Usually not; these statistics imply that the most likely time for a change of state to occur is "right now.")

A simulation model can represent more complex situations and interactions than a corresponding analytical model. However, it is inevitably more expensive and less flexible than the analytical model, and introduces simulation noise into the results (see Question 20). A simulation, moreover, masquerades under a cover of "realism"; it is easier to explain -- or at least *point to* -- elements of a simulation which "represent" real-world considerations. This use of real-world nouns *does not* mean that the simulation (or its data base) is any more realistic than the corresponding mathematical model.

Simulation framework compilers such as SIMSCRIPT or GPSS are powerful tools for cutting software costs, but if inexpertly used tend to introduce many hidden assumptions into the model: particularly assumptions that probability distributions are normal or uniform. The reader is also directed to Question 20 and the difficulty of obtaining valid statistics and sensitivities with a simulation.

These comments apply to pure software simulations as well as human-in-loop simulations (although in the latter case, the choice in favor of simulation may be forced by external needs).

11. <u>What are the trade-offs of the number of players in the model?</u>

The number of players in a model (whether real or simulated) can vary from one (against Nature) to two (one-on-one) to multiple groups (many-on-one or many-on-many), the last being typical of field experiments. The cost and time of preparing a verifiable simulation increases drastically (far faster than proportionately) with the number of players on each "side." The result (see Question 20) is a frequent paucity of validation and sensitivity data -- and of flexibility to accommodate new concepts in systems and operations.

On the other hand, a limited-player simulation does not ordinarily consider either the synergistic or the interfering effects of multiple players. Whichever approach was employed, the corresponding questions need to be asked. One reasonable solution is to perform most of the simulations at the one-one level and test the one-one hypotheses at a few scenario points with successively more complex many-many simulations. (See, however, Question 12.)

12. <u>Hierarchical models: something for nothing?</u>

An old concept recently being revived is the use of hierarchies of nested models to obtain both the fine-grained detail of a detailed simulation and the generic power of high-level models. In this concept, the fine-grained models are run first and their results used to form inputs to successively higher-level models -- this process continuing until the desired "macro" level of generality is obtained. It is potentially a good way of achieving a compromise between the two alternatives.

For reasons which by now seem obvious, the often-proposed case of a top-to-bottom model, which answers questions of broad generality while accepting inputs at a microscopic level of detail, poses just a few problems. While perhaps buildable, given enough human and machine time, it usually cannot be verified, validated, loaded, or run. The other extreme, that of building high-level models unconnected to lower-level data, frequently suffers from lack of real-world relevance and lack of sensitivity to the very parameters which may be vital to the manager's interests. The concept of a nested hierarchy of models, each feeding the next, has been viewed as a solution to this dilemma.

The reader will not be surprised that there is no such thing as a free lunch. The nested models have expenses and time budgets between the two extremes. They often do not consider the complex interactions and competing demands among organizations at a given level of detail (e.g., battalions) when proceeding to a higher level of generality (e.g., brigades). At best, they may allow at each stage for statistical variations among each set of subelements which aggregate to the next higher level of generality, but it is only with difficulty that they reveal the critical effects of synergy or of competition among these subelements. Indeed, a wing of aircraft is not a collection of individual aircraft.

Finally, the complexity of software necessary to successfully integrate such hierarchical models makes them necessarily less flexible to adapt to new tactics or systems not envisioned by their original designers.

13. What was left out?

Perhaps the most difficult evaluation incumbent on the manager or user of a modeling effort is to identify and assess the omissions (the *negative* assumptions of the work): those assumptions, frequently unstated, which rule out or omit considerations of importance. Such considerations might include material factors such as adverse weather, personnel costs, psychological characteristics, secondary logistics costs, or countermeasures (either present or developed in response to what is proposed); these are some frequent omissions. The omitted considerations might also include design options or operational options which ought, in retrospect, to have been considered: options such as alternative mobility vehicles, joint operations, reversal by preemption of the attacker-defender roles, alternative C^3I structures, or the use of product improvements built on existing hardware versus new developments.

What makes this issue particularly difficult is the fact that silent omissions by the model-maker are more often unconscious rather than deliberate. It is necessary to evoke them by asking "what-if" questions. Having evoked these missing elements (and there always are a few), the manager must now decide on their relevance. Were the choices acceptable?

14. To what extent is the model design-specific?

Every model is design-specific to some extent (that is, applicable to a specific class of problems, systems, or organizations whose characteristics are embedded in the algorithms). None is universal, despite enthusiastic claims to the contrary. The reader is referred to the famous Princeton final-examination question in Philosophy: "The Universe: Define and Give Two Examples."

The greater the extent to which a model is design-specific (i.e., the extent to which it addresses a specific problem or models a specific piece of hardware or organization), the greater the extent to which it can examine the details of what it models. More critically, however, any conclusions drawn from the model with respect to designs which were *not* incorporated are at best presumptive. Thus, an engagement model involving two opposing forces (aircraft, ships, armies) will have little to say regarding the intrusion of a third force with the rules of engagement of an armed neutral. Or a performance model of a tank will have little to say about a radical design incorporating robotic automation (or a "hero" type of unmanned vehicle).

This question is obviously closely related to Question 13. To assess the design limits of the model, it is necessary to ask creative questions of the form: "If we change System X to System Y or Solution X to Solution Y, will the model still be valid?"

15. How are human elements treated?

Most models are not man-in-the-loop models for a simple pair of reasons: lack of time and lack of budget. These constraints exhibit themselves, in practical terms, as a typical unavailability of appropriate, experienced, and critical players in sufficient quantity. Moreover, many models which *are* man-in-the-loop models should not be either: (1) because of the above impediments or (2) because of the difficulty of extracting generic findings from results involving specific human players with their human idiosyncracies.

Therefore, the difficulty of incorporating appropriately trained humans into a simulation (whether a war game or not) precludes their use except when their decision, invention, or psychomotor attributes are most critical. In those cases, moreover, the number of variables which can be independently studied with any statistical validity is very small. One common "solution" is to forego repeated runs (see Question 20) and to interpret the sparse data most imaginatively.

The human element is thus often replaced by submodels which are supposed to mimic human behavior. It is axiomatic that if the submodels do this by assuming "rational" or long-term optimal human behavior based on mathematical premises, they are not very realistic. There are three areas of endeavor in which these submodels have proven reliable, *provided* they have been verified against real-world data. These submodels relate to limited subsets of the following kinds of human activity:

(1) Psychomotor responses;

(2) Cognition and recognition; and

(3) Human decisions of a relatively straightforward "if-then" type, such as can be expressed by deterministic sequences.†

There are two other areas in which human performance submodels have some merit.

(4) Current work (at this time of writing) by researchers in artificial intelligence has achieved some success in modeling human decision behavior by techniques called *knowledge-based systems* or (a subset) *production rule systems*. These techniques represent fundamentally the efficient codification of complex decision procedures derived from observed human behavior. The reader should be cautioned, however, that such models only work for well-defined (but skilled and not necessarily deductive) decision processes where: (1) the objectives are clear; (2) the options are rigorously defined; and (3) wild-card options and any creative human action such as *ad hoc* deception are ruled out.

Again, probabilistic systems of decision behavior are allowed within this framework by some of the techniques which have found acceptance. The production rule methods (as distinguished from the more general knowledge-based category) are, however, Boolean and therefore suspect. Napoleon Bonaparte, in response to a question by his brother, wrote: "Above all, I expect my generals to be lucky." He expected them to make unprogrammed, novel decisions in the absence of full information and be "right." A good military decision maker with tactical authority knows when to throw away the book -- and the book contains those production rules.

(5) The final area of promise in human behavior modeling is just emerging from research: that of modeling the curve of risk acceptance or risk aversion exhibited by a decision maker. This amounts to the application of economic utility theory to psychological data, and some successful models have been built along these lines.

---

†*Note that the deterministic sequences may include purely logical assertions, fuzzy set elements, or ones based on probability evaluations (notably Bayesian models). They all are deterministic in the sense that a given set of inputs always produces the same output -- something which human decision makers perhaps do not* produce.

We suggest the reader regard with skepticism any software model purporting to represent human behavior unless: (1) it validly falls into one of the five categories described above; and (2) for that model, some validation is available.

We suggest the reader review any man-in-the-loop simulation with respect to the authenticity of the risk-reward environment, the capabilities and experience of the subjects, and the statistical validity of the work.

For those cases where problems relating to human elements remain on the table (and these cases are frequent), we suggest that the reader inspect the degree to which the analyst has assessed the sensitivities of the model to human behavior variations at some reasonable extremes. A jaundiced eye is suggested.

C. QUESTIONS DEALING WITH MODEL INTERNALS

16. How does the analyst deal with uncertainties?

In the insatiable quest for that important yet elusive quality known as "realism," model-builders often incorporate independent variables (i.e., inputs) about which not enough is known, *a priori*, to establish values or even distributions of values. The best that one can do under these circumstances is to perform sensitivity analyses and thereby establish "confidence limits" for the model's dependent variables (i.e., outputs). This practice is not frequently followed, however, because the model's results must then be reported with stated uncertainty -- an uncertainty which many model-builders believe (with some basis) will discredit the results in the eyes of their customers. Very often the only available solution to such output uncertainty (short of concealing it, ingenuously or disingenuously) is the recasting of the model and its outputs to minimize its sensitivity to variables unknown.

What frequently happens -- and for which the executive-user must be alert -- is the implicit or unconscious adoption of a train of reasoning of the following type:

> An input parameter of some type -- call it "X" -- is suggested, but there are no data for that variable. *Therefore*, it is a random variable.

> There are no data regarding the interaction of X with other inputs. *Therefore*, it is an independent random variable.
>
> There is no basis to assign a statistical distribution to X. *Therefore*, it is a uniformly distributed independent random variable.

Now this all sounds a bit disingenuous, but it is also ubiquitous. Moreover, many simulation languages *lend impetus* to this trap by adopting "default assumptions" modeled along the lines above. The danger of this common resolution is heightened by the realization that a linear combination of such random variables in a simulation causes the apparent variance of the results of successive runs to decrease rapidly (roughly as $1/\sqrt{N}$), lending a further air of credibility to what often amounts to a series of guesses.

The manager must be on guard for this form of resolution of the problem of unknown inputs; for as a general rule, it will result in an *understatement* of the uncertainty of the result by a substantial amount. He must inquire.

Instead of modeling unknown parameters as Monte Carlo variables, the analyst can:

(1) Make a determined effort to eliminate them by shedding attractive irrelevancies -- by alternative modeling and sensitivity reduction in the model itself.

(2) Determine the sensitivity of the outcome to variations in these remaining unknown parameters.

(3) Explore the effects of worst-case distributions and interdependencies if the uncertainty involves enemy tactical alternatives.

(4) Report the results as conditional uncertainties in the simulation product.

17. <u>What are the effects of hidden dependencies?</u>

This is not a highly technical question addressed, e.g., to the refinement of random number generators. Most programmers know how to write (or at least copy) good routines for this purpose. (There is one caveat, however. Most random number generators rarely work well if pressed for data at four times their standard deviations; if a model depends critically on events of this rarity, it had best not be a Monte Carlo model unless its perpetrator owns lots of computer stock.)

There are, however, many opportunities for hidden correlations to grow among the data in the model -- either because several variables are computed from the same common input or random event or because they are the products of strings of user input data which are really irrelevant to the problem but require laborious hand-coding. Such hand-coding of trivia (very common in some highly respected, large-scale models) is often left to clerks or overpaid engineers who, from sheer boredom, will often code the inputs in some regular fashion. This being a likelihood, the manager may be rewarded by some probing.

A third source: if mathematical techniques have been introduced to "stabilize" a simulation, their very presence can lead to the effects produced by correlation of data (i.e., increased confidence intervals).

To indicate the effects of non-independence of data, even a mild degree of pairwise correlation of data negates the effects of many runs for correlation as low as 0.6. To give the reader a "feel" for the situation: it has been observed that the correlation of data normally assumed independent in costing a military system is typically around 0.7, not zero. This is a statistician's inadequate explanation for the observed wide swings in program costs.

18. Was the model used within its range of legitimate applicability?

A frequent difficulty which arises in the application of a given model to a given military problem relates to the issue of its applicability to the situation of interest. This is frequently due to one or more of several factors:

(1) The model is invariably most useful in a predictive situation and may have underlying assumptions, which no longer apply, for studying the new situation.

(2) In the same new situation, the input data may stretch the model beyond its range of applicability.

(3) The impact of the results may reflexively make the model invalid.

This last factor comes about, e.g., in nuclear warfare models, in which after the first exchange neither the C^3 nor the logistic assumptions may apply.

The essential value of modeling lies in its use as a tool for extrapolation and prediction, and this use makes military modeling especially vulnerable to the effects of extension to environments, conditions, or situations in which the assumptions or data no longer apply.

The kinds of questions to ask are of the form: "Given the observed outputs, are the assumptions and input data still valid?" "Are the circumstances under which the model was used consistent with the range of circumstances under which it was tested?"

19. <u>What are the effects of the underlying mathematical assumptions?</u>

The assumptions employed in the model generally fall into three classes: those relating to the environment and operations, those relating to the systems and human elements, and the mathematical assumptions used to convert the real-world problem to a computationally tractable one. The first two have been discussed; the last is more difficult because the mathematical assumptions are not only quite abstract but also frequently unstated.

Some of these generic problems have been discussed in the context of simulations (Questions 10 and 16). Those relating to analytical techniques are more diverse and no less difficult to elucidate, complicated by the obvious desire to utilize "textbook" techniques as far as possible. These concern themselves particularly in the use of optimization techniques (e.g., linear programming, dynamic programming); in the use of integral approximations; and in the use of queueing, inventory, and other highly developed algorithmic techniques. The advantages of using analytical methods rather than simulations -- if the assumptions are acceptable -- are sufficiently profound to suggest their use.

More difficult are the unstated and implicit assumptions which result from the selection of any specific mathematical technique -- a selection which carries with it the baggage of the underlying assumptions of the mathematics itself. This problem is unavoidable since some mathematical model or another will always be used to quantify the results -- whether the model is cast as an analytical solution, a Monte Carlo simulation, or something else.

Certain models present special problems, e.g., the class of models which form a subset of linear programming (or its mathematical -- but not conceptual -- dual, zero-sum linear game theory). These problems are typical of those within a broader context of modeling where assumptions of linearity are frequently called into play. They include sensitivity analyses, factor analyses, and various algorithms used to combine variables (such as "measures of effectiveness"). Into this class fall multiplicative formulations as well as additive formulations since the former simply represent additions of logarithms.

Whenever linearization is employed (most particularly, in sensitivity analyses and "effectiveness" or "figure of merit" calculations), there will always be combinations of values which do not make sense yet which fall into the linear domain. That is the danger of linearization.

The questions to ask: What are the underlying mathematical models, what are their assumptions, how do they influence the results at the limits of application (maximum and minimum), and how do these assumptions constrain one's interpretation of the results?

20. How many iterations are needed?

There are two reasons for running a model many times over. The first of these relates to the use of any Monte Carlo method itself, where statistical techniques are used to sample from a very large space of possible input vectors (i.e., sets of input data) and the results statistically expressed. The second reason arises even with fully deterministic models (i.e., models whose outputs, whether expressed in "hard" numbers or statistical estimates, are the results of calculations and therefore not different for repeated runs). In these cases, nevertheless, experimental variation of input parameters is often employed to determine the robustness of the outcome and the sensitivity of the results to such variation.

One of the major failings of many Monte Carlo simulations is the lack of sufficient time or cost to run enough iterations of the model to allow any reasonable confidence in the outcome. Even if variation of the independent variables "of interest" (i.e., the variables of importance to the input of the model) is ignored, the number of required iterations is non-trivial and may lead to inadequate exercise of the model for sound results.

A special situation occurs when a hierarchical or nested Monte Carlo simulation is involved, in which an attempt is made to accommodate greater levels of detail by "nesting" a series of models at different levels of abstraction (see Questions 2 and 12), with each lower-level model output serving as input to the next higher level. This model architecture provides a subtle trap because the sequential staging of submodels itself can lead to underestimation of the uncertainties in the final result. The problem is a complex one and can best be unearthed by the interrogator asking questions which reveal that, for example, the modeling of a single flight of aircraft is not the same as the simultaneous modeling of N flights of aircraft in one squadron.

How many iterations are required by statistical considerations to determine each output variable, under each set of input conditions, to a given level of uncertainty or confidence? Were these many iterations carried out?

SOME CONCLUSIONS

We have tried to distill the experience of more years than the author and his colleagues care to count (and more tuition than they care to have paid) in analyzing, building, managing, and using models of many types -- some of which even fulfilled their charters. We have, therefore, suggested the context of a series of questions which the manager, the client, the user of the results of such models might ask, along with sufficient background to assist him or her in assessing the answers received -- the purpose of which is not to confer expertise but rather to provide a format for a dialogue, the results of which can be used to plan, guide, and manage the modeling effort and then assess the applicability of the model and its results to the problem at hand.

We have viewed the role of the client as related to that of the manager, and the role of the manager as properly one of a *mentor* who both directs and assesses the work, and have suggested 20 lines of inquiry to guide him. Knowing the difficulty of this task for the subject at hand and the opportunity for misapplication and inappropriate interpretation which it offers, we can only add: *Caveat Mentor.*

STEPHEN W. LEIBHOLZ

Stephen W. Leibholz is Editor of the Monographs of the Military Operations Research Society and President of Analytics. In the former capacity, he performs the various chores necessary to convert authors' manuscripts to books published by the Society, either of itself or in cooperation with a scholarly book publisher. The present volume is the fourth such publication.

As founder and CEO of Analytics, Mr. Leibholz supervises the operations of the company, providing analytical and engineering services primarily for the national security establishments as well as building information security, physical security, and communications systems for Government and industry.

Mr. Leibholz' career includes a tour as Manager of System Design and Analysis at the former Auerbach Corporation, where he learned operations research, and various other positions in defense industry and academe.

His professional areas of specialization include mathematical modeling of complex systems (he is the builder of several standard models in C^3I); the architecture and design of modern C^3I systems and hardware; special technical systems; information and communications security; and applications of statistical prediction techniques to, e.g., artificial intelligence.

Author and/or editor of some eight books, and author of several dozen scholarly papers and articles, Mr. Leibholz also serves on several Government committees. He is mentioned in American Men and Women of Science, several publications of Who's Who, and other compendia.

Index

AAW 27, 34

Abernathy, William 40

ABM Treaty 248

Ackoff, Russell L. 3

ADSTAP (Navy) program 194

Ad Hoc Committee on Professional Standards 10

Aircraft
  A-10 131, 139
  B-1 134, 227
  B-52 134, 139, 227
  F-15 139

Air defense 100-101, 121

Air Force Specialty Code (AFSC) 189

Air Training Command 197

Air warfare 281

Analog model 1-2

Analysis criterion modeling 282-283

Analytic models 42, 156, 165, 168-169, 336

Analytic scenarios 285

Index

Analytical representations 42

Antiaircraft artillery system 132

Approximation 315-317

Arms control 248-249

Arms Control and Disarmament Agency (ACDA) 32, 249, 251

Army Models Review Committee 7

Army Operational Research Group 72

Army Operations Research Symposium, 1969 71

Arrow, Kenneth 207

Artificial intelligence 17

Arunkumar, S. 208

ASM (Air-to-Surface Missile) 174-176

Assistant Secretary of Defense. *See also* systems analysis; assumptions in modeling 76, 285

ASW (Antisubmarine Warfare) 25, 27, 29, 112, 115, 171, 174, 178-179, 231, 286

Ballistic missile defense 101, 121

Barlow, R.E. 206, 208

Battilega, John A. 3

Battle planning 4, 23-24, 98

Battlefield Interdiction (BI) missions 140

Behavior. *See also* human resource model 187-198

Bellman, R. 206

Index

Bender, Edward 3

Berman, M.B. 213, 217

BIG DADDY Exercises 166

Blackett, P.M.S. 60, 66-69, 81, 83, 87, 295

Bondarenko, V.M. 17

Bonder, Seth 48, 295

Bortels, W.H. 84

"Bottleneck" transportation problem 212

Brewer, Garry D. 3, 20

Brothers, L.A. 76

Brown, Harold 231, 237

C^2; C^3; C^3I (Command and Control; Command, Control, and Communications; Command, Control, Communications, and Intelligence 5, 34, 97, 104, 110-111, 121-122, 141, 144, 237, 341, 347

CAPTOR mines 178

Cady, G.M. 213

Campaign model 181

Carrillo, M.J. 211

Center for Naval Analyses (CNA) 293-295, 299, 302, 310

Center for Naval Warfare Studies 167

Chuyev, Yu V. 17

Circular Error Probable (CEP) 142

Index

Clark, A.J. 211

Close Air Support (CAS) 140-141

Cohen, I.K. 207

Combat data 69-72

Computational plenty 58

Constant Elasticity of Substitution (CES) 193

Conway, R.W. 209

Cost-effectiveness 197, 225-226, 229-230

Cruise missile 101

Dantzig, G.B. 61, 212

Data banks 36-37, 71

Decision theory 6, 203, 205-206

Defense penetration effectiveness 101

Defense Planning System 168

Deitchman, S. 88

Deterministic approach 210

DoD Nuclear Warfare Analysis Center 123

Dorfman, Robert 57, 60, 63, 65, 70

Douhet, Giulio 95

Druzhinin, V.V. 17

Dynamic programming xii, 6

Index

Edison, Thomas A. 61

Effectiveness, Measure of 64-65, 74, 85, 254-256, 271, 282-283, 287, 348

Einstein, Albert 20

ELIM-COMPLIP (Army) system. *See also* transition model 194

Emerson, D.E. 217

English, John A. 84

Enthoven, Alain C. 79, 237, 241

"The Epistemological Problem" 303, 306, 309

Euclid 151

Ewell, J.J., LT GEN 79

Experiments 269-278

Explicit modeling 283-284

Falsifiability 337

Fermat, Pierre 300

Feyerabend, Paul 298, 308

Flax, Al 295

Force Mix Study 178

Force sizing 4, 23, 28-30, 34, 98, 165, 235-244, 247

Force-versus-force 134-144

Ford, L.R., Jr. 212

Freck, Peter G. 83

Fulkerson, D.R. 212

Index

Game theory 6

Garfinkel, R.S. 212

GLCM (Ground-Launched Cruise Missile) 121

Giblis, R.J. 207

Gingerich, Owen 1

Glaser, R.G. 207

Gödel, Kurt 305

Gomory, R.E. 213

Gorshkov, Sergei 17

Grange, Judith K. 3

Greene, William G. 71

Gross, A.J. 208

Hammer, P.L. 212

Hard Target Kill (HTK) 255

Hardison, David 271

Hayes, Robert 40

Heuristic algorithm 211

HF OTH-Backscatter radar 173

Hillestad, Richard J. 211, 217

Hitch, Charles J. 59, 64

Hoeber, Francis 3

Hopp, W.W. 209

Hu, M.R. 213

Human elements 342-344

Human resources planning 4, 23, 30-31, 98, 187-200

Hunt, I.A., Jr. 79

Hunter, L.C. 206

ICBM 102, 106, 112, 120, 141-142, 229, 236, 248, 287

Incompatibility of model hierarchies 35-36

Institute for Defense Analyses (IDA) 106

"Instructional Systems Development" 189

Jackson, Thomas J. (Stonewall) 304

Jacobson, S.E. 208

JCS Publication 1 42

JCS/SAGA ix, 42, 249, 281

Johnson, S.M. 209

Jones, R.V. 58, 69

Juncosa, M.L. 213

Kahn, Herman 59

Kalaba, R.E. 213

Kamins, M. 208

Kaufman, W. 308

Kendall, M.G. 12

Kent, Glenn 295

Kettelle, J.D., Jr. 208

Kimball, George E. 25-26, 57, 63-64, 68, 71, 75, 81-82, 84, 88, 296

Kontorov, D.S. 17

Koopman, Bernard O. 59, 80, 87

Kotkin, M. 211

Kuhn, Thomas 298

LANTCORTEX 166

Lag time 101-105, 107

Lagrange multipliers 211, 250

Lakatos, Imre 298

Lanchester equations 16, 38, 63, 87, 89, 143, 295

Lanchester, F.W. 87-88

Landi, D.M. 211

Leach, W. Barton 74-75

Least-cost allocation 208

Leckie, Robert 61

Lee, Robert E. 304

Level of Repair Analysis (LORA) 211

Levhari, D. 207

LITTLE DADDY Exercises 166

Index

Lientz, B.P. 213

Life cycle costs 226

Linear programming xii, 6-7, 72, 194, 203, 213, 338, 347

Lippiatt, T.F. 216

Logistics planning 4, 23, 31-32, 203-218

Low, Lawrence 3, 10

Maginot Line 319-320

Major Fleet Escort Study, U.S. Navy 1967 227

Manpower Mobilization System (NAMMOS) (Navy's) 192

Map gaming 100

Marginal analysis 211, 230

Markov model 338

Mastroberti, R. 212

Maxwell, W.L. 209

McLuhan, M. 14

McNamara, Robert S. 103, 235, 297

Meade, George G. 304

Measure of effectiveness (MOE) 64-65, 74, 79, 85, 254-256, 271

Measures of effectiveness, use of 64

Measure of merit 137, 142-143, 228-229

Michelson, Albert A. 309

MIDAS 212

MIRV 248, 253, 255

Mikhaylov, Yu. B. 17

Military Operations Research Society iii, 295

Miller, B.L. 209, 211

Miller, D.C. 308

Minuteman III system 229

Misappropriation of model results 36-37

Misappropriation of models 282-289

Miser, Hugh 70-71

MODIA 197-198

Models,
- AAW 34
- abstract 16
- ad hoc 5, 20, 26, 55, 151, 161, 297
- advanced penetration 106
- aggregation 127-137, 139-140, 189-190, 196-198
- air battle 150-152, 172
- air combat 189-190, 196-198, 296-297
- air defense penetration 96-101
- air-to-air combat 127-137, 139-141, 143
- air-to-ground combat 133, 135, 137-141
- analog 1-4
- analytic 42, 156, 165-169, 336
- arsenal exchange 101, 249-250
- assignment *See also* PRIDE/CLASP system (Navy); PROMIS/PJM system (Air Force); REQUEST/PIM system (Army) 187-189, 194-195, 197-198
- battalion-level combat 139
- battle 23-25
- CASTFOREM 162
- campaign 181
- *Catalog of Wargaming and Military Simulation Models* ix, 42, 281
- cobweb 16
- Code 50 249-250

Models (cont.)
combat 55-90, 127-137, 139-141, 143, 147, 155-156, 241-244, 271
Combined Arms Combat 147-149, 152-153, 155, 157, 162
command group training 148
communication 213
computer 68, 101, 103, 105, 107, 113-115, 123-124, 272-273, 281, 335-336
CORDIVEM 162
cost xii, 37
definition 1
descriptive 173
design 285
direct costing 261-263
division/corps evaluation 158
dynamic 194
Economic Order Quantity 6
effectiveness 231
engineering 148
evaluation 157-158
expected value 288
few-versus-few 126-127, 130-133, 142
force evaluation 158
FOREV 256-259
functional area 148
ground battle 140-141, 147-162, 319
ground forces 130-131
hierarchical 333, 340, 344-345
horizontal interaction, multiple 34-35
human performance 342-344
human resource 187-198
iconic 1
Improvement Program, Army 110
information warfare 5
interactive 167-168
inventory xii, 6, 203, 210-212, 227, 347
item system assessment 158
KRIEGSPIEL 160
land battle 147-162
limitations of 39-40
logistics 203-218
macro level 5, 14, 20, 35, 98-99, 101, 106-107, 110-111, 120, 127-128, 134-144, 153-154, 217, 236-237, 270, 333, 340
maintenance 206-209
management 148

Models (cont.)
manpower 188, 191-193, 195-198
many-versus-many 175-176
many-versus-one 175-176
Markovian 338
mathematical 2-3, 16, 42-49, 84, 206-208, 211, 269, 282, 296-298, 317-318, 321
medial level 147, 149-152, 159
micro unit engagement 5, 55-57, 79, 333
military 3-4, 6, 209
misappropriations of 282-289
missile effectiveness 142
multi-echelon inventory 210-212
multiple-unit engagement 5
Navy Resource 39, 243
one-sided nuclear 109-110
one-versus-one 127-135, 143
personnel 188, 192-198
phenomenological 25, 42-43, 170, 173-174
Poisson 338
predictions 316
predictive 6, 118
prescriptive 6-7, 34, 89
probability 25
qualitative 122
quantitative 115, 121-122
queueing 203
QUICK 249
real-time planning 158
Residual Cost 261
RPM 251, 253
scheduling 195-196
scientific 6, 10-12, 55-56
sea battle 166, 181-182
shop flow 208-209
simulation 3-5, 9, 37-38, 42, 84, 127-129, 132-133, 155-156, 175, 191,192, 203, 209, 211-212, 269, 338-340, 344-345
single mission 169
SIRNEM 251-253
standing 5, 20-22, 296-298, 332-333
static 193-194
steady state 193-194
stochastic 270, 276
strategic bomber penetration 95
strategic exchange 101, 109

Models (cont.)
  strategic missile exchange 128
  strategic operations 296
  structural assumptions 282
  structure 38-40, 318
  support evaluation 158, 210-211
  symbolic 1-2, 269, 271, 273, 278
  system engineering 149
  system preventative maintenance 206
  tactical engagement 175
  theater level 5, 14, 20, 36, 98-99, 101, 106-107, 110-111, 120, 127-128, 142-143, 153-154, 217, 236-237, 270, 333, 340
  training 42-43, 188, 195-198
  transition 194
  two-sided nuclear 109-110
  unit operations 147-149
  universal 2
  very aggregated 142-144
  weapon system performance 148
Modeling
  analysis criterion 282-283
  art of 40-41
  assumptions 285
  breadth versus level of detailing 283-284
  deterministic 283-284
  explicit 283-284
  implicit 283-284
  logistics infrastructure 203-218
  nuclear warfare 95-124, 215
  organizational 37-38
  parametric 283-284
  personality 37-38
  statistical 203

Monte Carlo simulation techniques 100, 156, 160, 177, 329, 346, 348-349

Morley, Edward W. 309

Morris, William T. 33

Morse, Philip M. 25-26, 57, 63-64, 68, 71, 75, 81-82, 84, 88, 296

Muckstadt, J.A. 211, 216-217

Index

National policy analysis 5, 23, 32-33, 247-265

National security analysis 328-329

National Security Council 251

NATO 98, 100, 108-110, 112, 117, 136, 138, 170, 295

Naval Enlisted Classification Code (NEC) 190

Naval War College 37, 167, 237

Navy Fighter Study 178

Navy Resource Model (NARM) 39, 243

Network theory x, 72, 203, 210, 213

Newman, James R. 62, 81

Newton, Isaac 308

Nolan, R.L. 212

Nuclear deterrence 249-250

Nuclear preemption 117

Nuclear weapon systems 283

Offensive counterforce capabilities 75-76

Office of Secretary of Defense 76-77, 232, 252

Operational limitations 111-112, 123

Operations, combat 169-170

Operations Research Society of America 10

Optimal replacement policies 206-207

Index

Paige, Emmet 332

Payne, Wilbur B. 15, 309

Peacetime analysis 55-58

Pickett, George E. 304

Planck, Max 309

Poincaré, Henri 41

Poisson processes 25, 37, 81

Political military gaming techniques 100, 109-110

Polya, George 307

PRIDE/CLASP system (Navy). *See also* assignment model 188, 194-198

PROMIS/PJM system (Air Force). *See also* assignment model 195

Prelaunch survivability 100-101, 106, 109, 115-116, 121

Presidential Review Memorandum No. 32 on Civil Defense (PRM-32) 253

President's Commission on Military Compensation 189, 194

Pritzker, A.A.B. 209

Prompt Hard Target Kill (PHTK) 255

Proschan, F. 208

Quade, E.S. 3, 22

Queueing techniques 203

Index

Radar equation 25

Rand Corporation 191, 243, 295-297

Rand Conference on Models, Games, and Simulations 295

Rao, M.R. 212

REQUEST/PIM system (Army). *See also* assignment model 195

Realism and relevance 8-10, 16, 18-19, 303, 344

Reliability 59, 203-204, 208, 259

Reproducibility 11, 29-30, 120, 241-242

Research games 156

RISOP/SIOP evaluation 249

SALT I 248, 251, 255

SALT (START) 32, 102, 249, 253, 255

SAM/ASM missiles 34, 128, 134-135, 138-139, 174-176

SAMSON 191

Sayers, Dorothy L. 12

Scarf, H. 211

Schaffer, M.B. 89

SEACONEX Exercises 166

Sea War 85 20

Search theory 63

Sensitivity 339-340, 344

SHAREM 166

Sherif, Y.S. 207

Sheshinski, E. 207

Shubick, Martin 3, 20

Silber, Jack 73

SIMSCRIPT compiler 339

SIRNEM model 251-253

SLBM 102, 106, 112, 255, 257

Smith, M.L. 207

Smith, Wayne K. 237, 241

Solandt, Omond 72, 81

Square law 88

SSBN 112, 115

SSN 174

Starkey, Lawrence 73

Stochastic demand 209-210

Strategic delivery system 140-141

Strategic deterrence 215

Strategic nuclear forces 236, 281, 287

Strategic strike force 134

Strategy and Force Evaluation (SAFE) 242-243

Supplemental Combat Information Report 71

Supply and Inventory Analysis 210-212

Surface-to-air combat 133

Survivability 111

Sweep rate 84-86

Systems analysis. *See also* Assistant Secretary of Defense 27, 76, 226

Tactical Warfare Simulation Program 297

Target acquisition systems 112

Taylor, James 3, 10

"The Teleological Problem" 303-306, 309

Theater force warfare 284-285

Theater nuclear delivery systems 110-111

Thomas, Clayton J. 25-26, 93

Tolstoy, Leo 328

TRASANA 21

TRIAD 284

Training games 156

Transparency 8, 30, 35, 119, 123, 228, 241-242, 252, 277-278

Transportation system design 204, 212-213

Uncertainty in arms control 253

"The Uncertainty Principle" of models 303, 309-310

UPTIDE 166

US Army Concepts Analysis Agency 271

Index

Validation and Verification (V&V) 48, 130, 315-324, 333-337, 339, 344

Veinot, A.F., Jr. 210

Volkov, A.F. 17

Wagner, H.M. 207

The War Game 8

Warfare,
- air 23, 281
- chemical 102, 107, 110-111, 114, 118
- conventional 102, 107, 110-111, 114, 118, 137, 236
- electronic 120-121
- general nuclear 23, 95-107, 110-112, 114, 117-118, 120-121, 141-142, 247
- land 23, 281
- Model Verification Conference 293
- naval 281
- sea battle 23
- theater force 284

War gaming 9, 24, 33, 35, 37, 42, 107, 111, 113, 117, 119-121, 159-160, 165-168, 242-244, 329, 336, 338

Warsaw Pact 110, 117, 170

Wartime applications 55-56, 97-98

Wartime operations 4, 23, 25-26, 324

Weapon allocation 250

Weapon performance 319

Weapon procurement 4, 23, 26-27, 34, 214, 225-232

Weapon selection 34, 36

Weapon systems 116, 204, 210-212, 215-216, 275, 316-317

Index

Weapon Systems Evaluation Group (WSEG) 83-84

Weiner, Milton 38, 108

Weirstrass, Karl 42

Welch, Jasper 17

Willsworth, N.E. 213

Witmore, W.F. 61, 64

Zagwill, W.I. 210

Zero sum game 229, 348